LABORATORY MANUAL FOR CHEMISTRY

LABORATORY MANUAL FOR CHEMISTRY

FOURTH EDITION

KAREN TIMBERLAKE

LOS ANGELES VALLEY COLLEGE

HarperCollins*Publishers*

Cover illustration: Baby Shampoo Film, Photo Researchers, Inc. © Tom Branch/
Science Source

Laboratory Manual to accompany CHEMISTRY, Fourth Edition

Copyright © 1988 by HarperCollins*Publishers* Inc.

ISBN: 0-06-046689-8

90 91 92 93 9 8 7 6 5 4

TABLE OF CONTENTS

Preface
Introduction

PREFACE

THE LABORATORY

Here you are in a chemistry laboratory with your laboratory book in front
of you. Perhaps you have already been assigned a laboratory drawer, full
of glassware and equipment you have never seen before. Looking around you
may see bottles of chemical compounds, balances, burners and other
equipment that you are going to use. This may very well be your first
experience with laboratory equipment and experimental procedures. At this
point you may have some trepidation about what is expected of you. This
laboratory manual is written with those considerations in mind.

USING THE LABORATORY MANUAL

Each experiment begins with objectives. These tell you the concepts you
will be studying in that experiment. Each experiment is keyed to the
objectives in the text for your study reference. You will also find a
list of the materials needed for that particular experiment.

The procedures are written to guide you through the experiment. When
you are ready to begin your experiment, remove the laboratory record sheet.
As you follow the procedures for the experimental activities, record your
answers and data on the record sheet. Complete the required calculations.
The follow-up questions test your understanding of the concepts from the
experiment as well as solving problems related to the experiment.

INTRODUCTION

PURPOSE OF THE LABORATORY

It is important to realize that the value of the laboratory experience depends on the time and effort you invest in it. Only then will you find that the laboratory can be a valuable learning experience and an integral part of the chemistry class. The laboratory gives you an opportunity to go beyond the lectures and words in your textbook and experience the experimental world of trial and error from which conclusions and theories concerning chemical behavior and properties are drawn. In some experiments the chemical concepts are correlated with health and biological concepts. Chemistry is not an inanimate science, but one that helps us to understand the behavior of living systems.

It is my hope that the laboratory experience will help illuminate the concepts you are learning in the classroom. This experimental experience can help make chemistry a real and exciting part of your life and provide you with skills necessary for your future.

THE LABORATORY WITH ITS EQUIPMENT, GLASSWARE AND CHEMICALS HAS THE POTENTIAL FOR ACCIDENTS. IN ORDER TO AVOID DANGEROUS ACCIDENTS OR TO MINIMIZE THEIR DAMAGE, PRECAUTIONS MUST BE TAKEN BY THE STUDENT AND TEACHER TO INSURE THE SAFETY OF EVERYONE IN THE LABORATORY. APPROVED EYE PROTECTION IS REQUIRED FOR ALL EXPERIMENTAL PROCEDURES.

PREPARATION FOR LABWORK

1. Before you start an experiment, read the entire section on the experiment you are about to perform. If you have been given a laboratory schedule, read the experiment before you come to the laboratory. Make sure you fully understand the experiment before starting the actual work. If you are in doubt, ask your instructor to clarify the procedures.

2. Do only the experiments that have been assigned by your instructor. No unauthorized experiments are to be carried out in the laboratory. Do your experiments at the assigned times in the lab. Do not work alone in a laboratory.

3. Wear sensible clothing in the laboratory. Loose sleeves, short pants or open-toed shoes can be dangerous. Tie back long hair to put it out of reach of the flame from a bunsen burner and from chemicals. You may wish to wear protective clothing such as an apron or a lab coat.

4. No food, drinks or smoking is allowed in the laboratory. Wash your hands before you leave the laboratory. Do not let your friends visit in the laboratory; have them wait outside.

5. Clear the laboratory counter of all your personal items such as backpacks, books, sweaters and coats. Find a storage place in the lab for them. All you need is your laboratory manual, calculator, text and the equipment provided in the lab.

HANDLING CHEMICALS

1. To avoid contamination of the chemical reagents, never insert droppers, pipets or spatulas into the reagent bottles. DOUBLE CHECK the label on the reagent bottle carefully before you remove a chemical. Pour or transfer a chemical into a small container (beaker, flask, etc.) of your own. Take only the quantity of chemical necessary for your particular experiment. Make sure you cover the reagent bottle with its original cover. Return the reagent bottles to their proper location in the laboratory; do not keep them at your desk. Label the compounds in each container. If you forget, discard the contents and do not try to use it in an experiment.

2. Never return unused chemicals to the reagent bottles. Liquids and water-soluble chemicals may be washed down the sink with plenty of running water. Organic liquids should be disposed of in special containers in the hoods. Dispose of solid chemicals in the special trash cans in the laboratory. Do not wash solids down the drain.

3. When diluting acids, always pour ACID INTO WATER.

HEATING CHEMICALS

1. Only glassware marked Pyrex or Kimax can be heated. Other glassware will shatter when heated. When heating solids or liquids in a test tube over an open flame, never point the open end at anyone, or look directly into it. Hold the test tube in a test tube holder at an angle- not upright - over the flame, heating the sides and bottom while you continuously move the test tube.

2. Never heat a flammable liquid over an open flame.

3. Be careful of picking up equipment you may just have been heating. This might be an iron ring, a clay triangle, a test tube, a beaker, a crucible or a flask. A heated piece of iron or glass looks no different than a piece at room temperature.

4. Do not place any heated object on the balance. Let the object cool first on a heat pad at your table.

TASTING AND SMELLING CHEMICALS

1. NEVER taste a chemical.

2. When required to note the odor of a chemical, first take a deep breath of fresh air and exhale normally. Then with your hand waft the vapors toward your nose and sniff very gently. Do not inhale the fumes directly.

LABORATORY CLEANUP

1. Start your cleanup 15 minutes before the laboratory ends. Clean and dry equipment. Return any borrowed equipment to the stockroom. Make sure the gas and water at your table are turned off. Clean and dry your work area.

2. Check the balance you used. Clean up any spills in the area.

EMERGENCY PROCEDURES

1. Learn the location and use of the emergency eye-wash fountain, the emergency shower, fire blanket, and fire extinguishers. Memorize their locations in the laboratory.

2. If a chemical is spilled on the skin, IMMEDIATELY FLOOD THE AREA WITH WATER. If a chemical gets into the eyes, assistance will be needed to flood the eyes with water at the eye-wash fountain. In both instances, water should be continuously applied for ten minutes or more. Notify your instructor when any chemical has spilled on your skin or gotten in your eyes. Do not try to neutralize an acid or a base spill.

3. Clean up spills at your table or floor immediately. An absorbent compound may be available to soak up the chemical. It can then be swept up with a brush and pan. Broken glassware is also immediately swept up with a brush and pan. Notify your instructor of any mercury spills. Mercury spills need special attention.

4. If clothing or hair catch on fire, get the student to the shower or use the fire blanket. Cold water or ice can be applied to small burns. Major burns require immediate medical attention. If a chemical fire is involved, use a fire extinguisher to douse the flames. Shut off gas burners in the laboratory.

SAFETY QUIZ

1. Approved eye protection is to be worn

    a. for certain experiments.
    b. only for hazardous experiments.
    c. all the time.

2. Eating in the laboratory

   a. is not permitted.
   b. is allowed between noon and 1 p.m.
   c. is all right if you're careful.

3. If you need to smell the odor of a chemical, you should

   a. inhale over the test tube.
   b. take a breath of air, exhale, and fan the chemical
      vapors toward you.
   c. pour some in your hand and smell it.

4. When heating liquids in a test tube, you should

   a. move the tube back and forth through the flame.
   b. look directly into the open end of the test tube
      to see what is happening.
   c. hold it tightly in your hand so it won't fall
      and break.

5. Unauthorized experiments

   a. are all right as long as they don't seem hazardous.
   b. are all right as long as no one finds out.
   c. are not allowed.

6. If a chemical is spilled on the skin you should

   a. wait to see if it stings.
   b. flood with lots of tap water.
   c. add another chemical to absorb it.

7. When taking liquids from a reagent bottle

   a. use only a clean dropper.
   b. pour the reagent into a small container.
   c. put back what you don't use.

8. In the laboratory, open-toed shoes and shorts

   a. are dangerous and should not be worn.
   b. are OK if the weather is hot.
   c. are all right if you wear a lab apron.

9. When is it all right to taste a chemical?

   a. NEVER
   b. When the chemical is not hazardous
   c. When you use a clean beaker

10. After you use a reagent bottle,

    a. keep it at your desk in case you need more.
    b. return it to its proper location.
    c. play a joke on your friends and hide it.

11. Before starting an experiment,

    a. read the entire procedure.
    b. ask your lab partner how to do the experiment.
    c. skip to the laboratory record and try to figure
       out what to do.

12. Working alone in the laboratory

    a. is all right if the experiment is not too hazardous.
    b. is not allowed.
    c. is allowed if you are sure you can complete the
       experiment without help.

13. You should wash your hands

    a. before you leave the lab.
    b. only if they are dirty.
    c. before eating lunch in the lab.

14. Personal items (books, coats, etc.)

    a. should be kept on your lab bench.
    b. should be stored out of the way, not on the
       lab bench.
    c. left outside.

15. When you have taken too much of a chemical you should

    a. return it to the reagent bottle.
    b. store it in your lab locker for future use.
    c. discard it using proper disposal procedures.

16. In lab, you should wear

    a. sensible, protective clothing.
    b. something fashionable.
    c. shorts and loose sleeve shirts.

17. When you dilute acids,

    a. add the acid to water.
    b. add the water to the acid.
    c. either way is all right.

18. If a chemical is spilled on the table,

    a. clean it up right away.
    b. let the stockroom help clean it up.
    c. forget about it.

19. If mercury is spilled,

    a. pick it up with a dropper.
    b. call your instructor - mercury spills need
       special attention.
    c. push it under the table where no one can see it.

20. If your hair or shirt catches on fire, you should

      a. run to the nearest exit.
      b. let it burn out.
      c. use the shower or fire blanket to put it out.

ANSWERS TO SAFETY QUIZ

1. c	6. b	11. a	16. a
2. a	7. b	12. b	17. a
3. b	8. a	13. a	18. a
4. a	9. a	14. b	19. b
5. c	10. b	15. c	20. c

LABORATORY EQUIPMENT

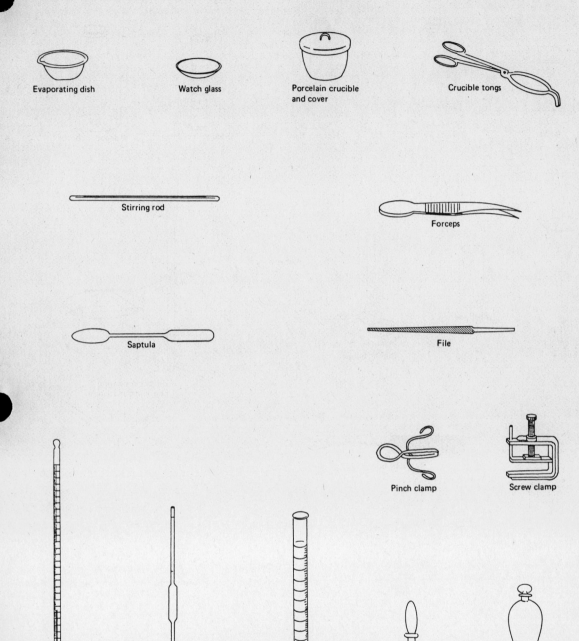

Evaporating dish

Watch glass

Porcelain crucible and cover

Crucible tongs

Stirring rod

Forceps

Saptula

File

Pinch clamp

Screw clamp

Thermometer

Pipet

Buret

Eyedropper

Separatory funnel

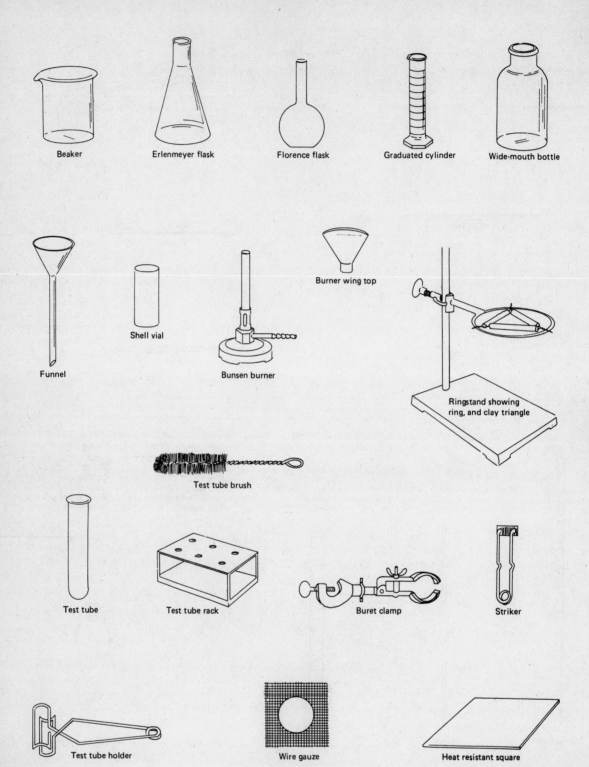

Beaker

Erlenmeyer flask

Florence flask

Graduated cylinder

Wide-mouth bottle

Funnel

Shell vial

Bunsen burner

Burner wing top

Ringstand showing ring, and clay triangle

Test tube brush

Test tube

Test tube rack

Buret clamp

Striker

Test tube holder

Wire gauze

Heat resistant square

EXPERIMENT 1   MATH AND THE CALCULATOR

PURPOSE

1. State the correct number of significant figures in a measurement.
2. Round off a calculated answer to the correct number of significant figures.
3. Perform mathematical calculations and give final answers with the correct number of significant figures.
4. Write a number in scientific notation.

MATERIALS

scientific calculator
pencil or pen

KEYED OBJECTIVES IN TEST: Appendix A and B

DISCUSSION OF EXPERIMENT

In the sciences, we make many measurements such as the mass or length of an object. The values obtained from measurements are called **measured numbers.** It is important that you use these measured numbers correctly in calculations and that you report calculated answers properly.

LABORATORY ACTIVITIES

A. Significant Figures

All the figures in a measured number are called **significant figures.** The following rules can be used to determine the number of significant figures in a measured number:

Significant figures are	Example	Number of Significant Figures
a. all nonzero digits.	455.2 cm	4
	4.52 g	3
b. zeros between or after other digits in a decimal number.	305 m	3
	50.20 L	4
	4.0 kg	2

Zeros are <u>not</u> significant if they occur

c. in front of a number.	0.0005 lb	1
d. in large numbers that have no decimal point.	25,000 ft	2

Now complete Section A on your **LABORATORY RECORD** which can be found at the end of this experiment.

B. Rounding Off

   After you have obtained some numbers from measurements, you may
begin to carry out mathematical operations such as multiplication,
division, addition or subtraction.   The results obtained from such
operations are called **calculated answers**.   When a calculated answer
has too many numbers, it is necessary to round off the answer.

Rules for Rounding Off

If the first number to be dropped is

   a. less than 5, it and all following numbers are simply dropped.

   b. 5 or greater, the preceding digit is increased by 1.

		3 sig figures	2 sig figures
Example:	75.6243 rounds off to	75.6	76
	1.528392 rounds off to	1.53	1.5

Now complete Section B in the LABORATORY RECORD.

C. Multiplication and Division

   When a calculation uses multiplication and/or division, the final
answer can have only as many significant figures as the measured
number with the fewest significant figures.

Example 1-1:   3.56   x   4.9

Solution: On the calculator you would

   Enter 3.56   Press   X   Enter   4.9   Press   =

The calculator will show the following numbers:

         17.444   (calculator answer)

The calculated answer has too many figures. It must be rounded off to
two significant figures since the measurement 4.9 in our calculation
has just two significant figures.

      Correct answer: 17

Example 1-2:
$$\frac{(0.025)(4.62)}{3.44}$$

Solution:

On your calculator, the values are entered and the correct operation key is pressed. Note that you do not need to write down any intermediate answers that appear. You need only record the answer that appears after all the mathematical operations are done.

enter 0.025   press X   enter 4.62   press ÷   enter 3.44   press =

= 0.0335755   (calculator answer)

= 0.034   final answer rounded off to two significant figures, since 0.025 has only two significant figures

Complete Section C in the LABORATORY RECORD.

D.   Addition and Subtraction

In calculations using addition and/or subtraction, the final answer can have the same number of digits as the measurement with the fewest digits after the decimal point.

Example 1-3:   Add:

42.11
4.056
30.1   (only 1 digit after the decimal point)

76.266   calculator answer

76.3   rounded off answer with 1 digit past the decimal point

Example 1-4:
Subtract: 14.621
- 3.39   2 digits past the decimal point

10.231   calculator answer

10.23   rounded off answer with 2 digits past the decimal point

Now complete Section D in the LABORATORY RECORD.

3

E. Scientific Notation

In scientific work, you will use some very small numbers such as 0.000000025 m and some very large numbers such as 4,000,000 g. It is common for a scientist to express large and small numbers in terms of a coefficient and a power of 10.

$$0.00001 = 10^{-5} \qquad\qquad 10,000 = 10^4$$

$$0.001 = 10^{-3} \qquad\qquad 1,000 = 10^3$$

$$0.01 = 10^{-2} \qquad\qquad 100 = 10^2$$

$$0.1 = 10^{-1} \qquad\qquad 10 = 10^1$$

To write numbers in scientific notation, use the following rules:

For large numbers,

a. move the decimal point to the left until it is placed just after the first digit in the number. This is the coefficient.

b. state a power of ten that is equal to the number of places the decimal point moved.

Example: $\quad 35,000 \quad = \quad \underset{\text{coefficient}}{3.5} \quad \text{x} \quad \underset{\text{power of ten}}{10^4}$

$$2,000,000 \quad = \quad 2 \quad \text{x} \quad 10^6$$

For small numbers,

a. move the decimal point to the right until it is placed just after the first digit in the number.

b. state a negative power of ten that is equal to the number of places the decimal point moved.

Example: $\quad 0.0042 \quad = \quad 4.2 \text{ x } 10^{-3}$

$$0.0000815 \quad = \quad 8.15 \text{ x } 10^{-5}$$

Now complete Section E in the LABORATORY RECORD.

EXPERIMENT 1   MATH AND THE CALCULATOR

LABORATORY RECORD   NAME_____
                                           DATE_____
                                           SECTION_____

A. Significant Figures

   State the number of significant figures in each of the
   following measured quantities:

   a.   4.5 m _____   d.   204.52 g _____

   b.   0.0004 L _____   e.   625,000 mm _____

   c.   805 lb _____   f.   1.0065 km _____

B. Rounding Off

    Round off each of the following numbers to the number of
    significant figures indicated:

                              Number of Significant Figures

                        three figures   two figures

    a.   49.34   _____   _____

    b.   5.448   _____   _____

    c.   85.83423   _____   _____

    d.   532,800   _____   _____

    e.   143.63212   _____   _____

C. Multiplication and Division

    Perform the following multiplication and division
    calculations. Give final answers with the correct number
    of significant figures:

    a.   4.5   x 0.28   =   _____

    b.   0.1184 x 8.0 x 0.034   =   _____

    c.   11.4   =   _____
          ____
          2.3

    d.   (42.4)(5.6)   =   _____
         _____
            1.5

    e.   (35.56)(1.45)   =   _____
         _____
          (4.8)(0.56)

5

D. Addition and Subtraction

Perform the following addition and subtraction calculations. Give final answers with the correct number of significant figures.

a. 6.25 g   + 0.683 g   = _____

b. 13.45 mL + 6.5 mL + 0.4552 mL = _____

c. 145.5 m   + 86.58 m + 1045 m   = _____

d. 245.625 kg - 80.2 kg   = _____

e. 4.62 cm   -   0.885 cm   = _____

E. Scientific Notation

Write the following numbers in scientific notation:

a. 4,450,000   _____

b.   38,000   _____

c. 25.2   _____

d. 0.00032   _____

e. 0.0000000021   _____

Write the following as ordinary numbers:

a. 4×10^2 _____

b. 5.5×10^4 _____

c. 1.88×10^6 _____

d. 8×10^{-3} _____

e. 4.25×10^{-5} _____

QUESTIONS AND PROBLEMS

1. A number that counts something is an exact or pure number. When
    you say 4 cups, or 5 books or 2 watches, you are using exact
    numbers. However, when you use a ruler to measure the height of
    your friend as 155.2 cm, you obtain a measured number. Why are
    some numbers called exact numbers while other numbers are called
    measured numbers?

2. Bill and Beverly have measured the sides of a rectangle and
    recorded that the length of the rectangle is 6.7 cm and the
    width is 3.9 cm. When Bill calculates the area by multiplying
    the length by the width, he records an answer of 26.13 cm^2.
    However, Beverly records an answer of 26 cm^2 for the area.

    a. Why is there a difference between the two calculated answers
        when they used the same measurements?

    b. You are going to tutor Bill. What would you tell him to help
        him correct his answer?

EXPERIMENT 2   THE METRIC SYSTEM: MEASURING LENGTH

PURPOSE

1. To read metric measurements for length accurately.
2. To record measurements for length correctly.
3. To determine experimentally U.S.-metric conversion factors.
4. To compare an experimentally determined factor for length
    to the accepted value of the factor.

MATERIALS

meterstick or ruler
string

KEYED OBJECTIVES IN TEXT: 1-1, 1-2, 1-3, 1-4, 1-5

DISCUSSION OF THE EXPERIMENT

Scientists and allied health personnel must be able to understand
and carry out laboratory procedures, take measurements, read
thermometers, and report results accurately and clearly. How
well they do these things can mean life or death to a patient.

        The system of measurement used in science, and frequently
used in hospitals and clinics, is the metric system, which is a
decimal system (based on units of 10). The unit of length in the
metric system is the meter (m). Using an appropriate prefix, you
can indicate a length that is greater or less than a meter.
Table 2-1 lists some of the most commonly used metric units of
length.

Table 2-1   Some Metric Units Used to Measure Length

kilometer(km)	1 km = 1000 m	
meter(m)	1 m	
decimeter (dm)	0.1 m (1/10 m)	1 m = 10 dm
centimeter (cm)	0.01 m (1/100 m)	1 m = 100 cm
millimeter (mm)	0.001 m (1/1000) m	1 m = 1000 mm

EXPERIMENT 2 THE METRIC SYSTEM: MEASURING LENGTH

LABORATORY ACTIVITIES

PART I: METRIC UNITS

A. THE METERSTICK

Obtain a meterstick or ruler. Observe the markings on the
meterstick. (See Figure 2-1) Answer the questions on the
laboratory record.

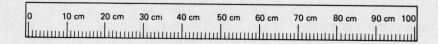

Figure 2-1   A meterstick.

B. ESTIMATIONS AND MEASUREMENT

Now try some estimations using metric units. An estimation is a
guess about a length without actually making a measurement.
Give estimations in centimeters(cm) for the length of your little
finger, the length from your elbow to your wrist, the length of
your foot, the distance around your head, and the distance around
your wrist. Be sure that you include the unit of measurement for
each estimation. A number alone does not give sufficient
information.

When you have completed your estimations, use a meterstick
and, the piece of string to measure the lengths in centimeters
(cm). Measure as precisely as you can. The last figure in your
measured number will be an estimation. DO NOT TRY TO ROUND OFF
NUMBERS FROM MEASUREMENTS. Record these measurements next to the
estimations you made.

EXPERIMENT 2   THE METRIC SYSTEM: MEASURING LENGTH

PART II: U.S.-METRIC EQUALITIES

A measurement of length can be expressed in the U.S. customary system as well as in the metric system. When we know both, we can make a comparsion of the two units of measurement in the form of a fraction. This comparison is called a **conversion factor**. For example, we find that the time of 3.0 hours is equal to the time of 180 minutes. We can write this equality as a fraction and divide by the denominator. As a result, we obtain a conversion factor for minutes and hours.

EQUALITY   3.0 hr = 180 min

FRACTION   CONVERSION FACTORS

$$\frac{180 \text{ min}}{3.0 \text{ hr}} \quad = \quad \frac{60 \text{ min}}{1.0 \text{ hr}} \quad \text{or} \quad \frac{1.0 \text{ hr}}{60 \text{ min}}$$

In this section, we will experimentally determine the U.S.-metric conversion factor for centimeters and inches. (See Figure 2-2.)

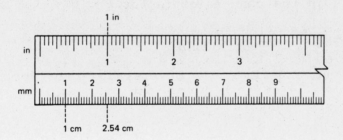

Figure 2-2   The relationship between the U.S. and metric systems for measuring length.

C. COMPARING CENTIMETERS AND INCHES

C-1 Use the meterstick or ruler to measure the length (vertical distance) of this page in cm. Record.

C-2 Use the meterstick or ruler to measure the length of the page in inches.   Record.

Note: Express any fraction of an inch or centimeter as a decimal.

Measurement with fraction	Decimal form
4 1/2 in.	4.5 in.
10 3/4 in.	10.75 in.

Also, do not mix units. For example, a measurement of 8 cm and 4 mm is 8.4 cm.

C-3 Now you are ready to calculate a conversion factor using your measurements. Form a fraction by placing the length you measured in cm in the numerator and the length in inches in the denominator. Divide by the denominator to obtain your experimental conversion factor. Be sure to give the correct number of significant figures in your final answer and to keep both units in the conversion factor.

$$cm/in \quad = \quad \frac{length(cm)}{length(in.)}$$

C-4 Compare your experimental conversion factor (cm/in) with the given value(2.54 cm/in.) Why might your experimental value be somewhat different?

D. MEASURING YOUR HEIGHT IN CENTIMETERS

D-1 Measure your height in centimeters. Record.

D-2 Measure your height in inches. If you are using the inch scale on a meter stick, note that there are more than 39 inches in its length.   Record.

D-3 Starting with the value of your height in inches, use the given conversion factor to calculate your height in centimeters. Show the proper setup for your calculations on the record sheet.

$$height (cm) \quad = \quad height (inch) \quad x \quad \frac{2.54\ cm}{1\ inch}$$

Compare your measured height with the value you calculated for your height. Explain why they may be close or not close.

EXPERIMENT 2   THE METRIC SYSTEM: MEASURING LENGTH

LABORATORY RECORD

NAME_____
DATE_____
SECTION_____

PART I: METRIC UNITS

A.   <u>METERSTICK</u>

What are the units represented by the <u>numbers</u> on the meterstick?

What do the smallest units marked on the meterstick represent?

Complete the following statements:

There are _____cm in 1 m.

There are _____mm in 1 m.

There are _____mm in 1 cm.

B. <u>ESTIMATIONS</u> <u>AND</u> <u>MEASUREMENT</u>

Length	Estimation (cm)	Measurement(cm)
little finger	_____	_____
elbow to wrist	_____	_____
around wrist	_____	_____
foot	_____	_____
around head	_____	_____

EXPERIMENT 2   THE METRIC SYSTEM: MEASURING LENGTH

PART II: U.S. – METRIC EQUALITIES   NAME_____

C. <u>CENTIMETERS</u> <u>AND</u> <u>INCHES</u>

C-1   Measured length of page in cm   _____

C-2   Measured length of page in inches   _____

C-3   Calculated conversion factor

$$\frac{\rule{3cm}{0.4pt}\ \text{cm}}{\text{inch}} \quad = \quad \frac{\rule{3cm}{0.4pt}\ \text{cm}}{\text{inch}}$$

C-4   How does your experimental conversion factor compare to the
       given value of 2.54 cm/inch?

Give some reasons that might explain your answer above.

D. <u>HEIGHT</u> <u>IN</u> <u>CENTIMETERS</u>

D-1   Height in centimeters(measured)   _____

D-2   Height in inches(measured)   _____

D-3   Calculated height in centimeters   _____
       Calculations:

How does your measured height compare to the calculated
height?

Give some reasons for your answer above.

EXPERIMENT 2   THE METRIC SYSTEM: MEASURING LENGTH

QUESTIONS AND PROBLEMS   NAME_____

1. Give one advantage of using the metric system.

2. Measure the following lengths on the line in millimeters. Convert the measurements to the other units listed:

A _____B _____C

Distance	mm	cm	m
AB	_____	_____	_____
BC	_____	_____	_____
AC	_____	_____	_____

3. Using conversion factors, set up and solve the following problems. Show your work. All numbers must have units and units must cancel. Give final answers with the correct number of significant figures.

   a. A pencil has a length of 0.085 m. Calculate its length in millimeters(mm).

   b. A newborn infant has a length of 20.2 inches. What is the newborn's length in cm?

   c. A roll of gauze measures 1250 mm. What is the length of the gauze in feet?

   d. Plastic tubing sells for 35 cents/ft. What is the cost in dollars of 275 cm of plastic tubing.

15

EXPERIMENT 3   THE METRIC SYSTEM: MEASURING VOLUME

PURPOSE

1.   To read and record volume measurements accurately.
2.   To calculate U.S.-metric conversion factors for volume.
3.   To compare an experimentally determined factor with the accepted value.
4.   To measure the volume of a solid directly, and by volume displacement.

MATERIALS

50-mL graduated cylinder
1-L graduated cylinder(stockroom)
1-qt measuring cup (or 1-cup or 1-pint measure)
a set of partially filled graduated cylinders marked 1,2, and 3
a wood or metal solid with a regular shape
large graduated cylinders (100, 250, or 500-mL), to fit solid

KEYED OBJECTIVES IN TEXT: 1-1, 1-2, 1-3, 1-4, 1-5

DISCUSSION OF EXPERIMENT

The volume of a substance measures the space it occupies. In the metric system, the unit for measuring volume is the liter. Smaller volumes are often expressed in deciliters(dL), or milliliters (mL).

liter (L)	1 L	
deciliter (dL)	0.1 L	1 L = 10 dL
milliliter (mL)	0.001 L	1 L = 1000 mL

Note: One cubic centimeter (cm^3 or cc) is equal to 1 mL. The terms are used interchangeably.

$$1 \ cm^3 \ = \ 1 \ cc = \ 1 \ mL$$

LABORATORY ACTIVITIES

PART I: READING THE VOLUME OF A FLUID IN A GRADUATED CYLINDER

A. VOLUME OF A LIQUID

Reading a Graduated Cylinder
The graduated cylinder is used to measure the volume of liquids in the laboratory. To read a graduated cylinder properly, set the cylinder on a level surface and bring you eyes even with the liquid level inside.   Read the volume measurement at the lowest point of the meniscus. The **meniscus** is the curve at the top of the fluid layer that is lower at the center. Be sure to read the same point of the meniscus in each volume measurement. See Figure 3-1.

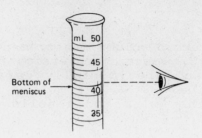

Figure 3-1   A graduated cylinder. The bottom of the meniscus
              is read as 42.0 mL.

You will need to determine what the lines or divisions on the
graduated cylinder represent.   On a 50-mL graduated cylinder, each
line might measure 1.0 mL.   However, on a 250-mL cylinder, each
line may represent 5.0 mL, while on a 1000 mL cylinder, each line
would be 10 mL.

To determine the volume represented by one division, count
the number of divisions between two lines that show the volume
measurement.   For example, there are ten divisions between the 300
mL and the 400 mL markings on a large graduated cylinder.   That
means that a volume of 100 mL is divided into 10 sections or that
there are 10 mL per division.

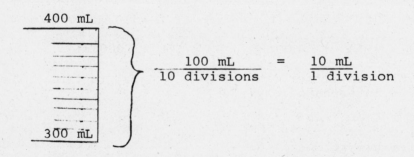

Read the volumes of liquids that are in the graduated cylinders.
Record the volumes in milliliters.   State the volume in liters.

EXPERIMENT 3   THE METRIC SYSTEM: MEASURING VOLUME

B. QUARTS AND MILLLITERS

A volume of water can be measured in both metric and English units.
In this experiment, we will compare the volume of a quart with its
volume in milliliters.   From the comparison, we can derive a
conversion factor.

B-1   Fill a measuring cup with water to the 1-qt mark. (Or use a 1-
      pint measure twice, or a 1-cup measure four times.) Transfer
      the water to the 1-L graduated cylinder. Record the number
      of milliliters in 1 qt.

B-2   Calculate the experimental conversion factor for mL/qt.

$$\frac{mL}{1\ qt} \qquad \text{CONVERSION FACTOR}$$

B-3   How does your experimental conversion factor for mL/qt
      compare with the given value, 946 mL/1qt? Explain why your
      results might be somewhat different?

PART II: VOLUME OF A SOLID

C.   VOLUME BY DIRECT MEASUREMENT

The volume of a regularly shaped object can be determined by direct
measurement.   For rectangular solids, the length (l), width (w)
and height (h) are measured. If the solid is a cylinder, you need
to measure the diameter(d) and the height (h). The measurements of
the solid are then placed in a formula for the proper geometrical
shape to calculate the volume.

C-1   Obtain a solid that has a regular shape such as a wood block,
      metal cube, rectangular solid, or cylinder. Keep this solid
      for use in Part D of this experiment.

C-2   Use a metric ruler to determine the dimensions in
      millimeters(mm) and centimeters(cm) of the solid.

      Dimensions needed:

         cube, measure length(L) of a side

            $V = L^3$

         rectangular solid, measure length, width and height

            $V = L \times W \times H$

            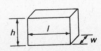

19

cylinder,   measure length and radius (r = 1/2 diameter)

$V = 3.14 \times r^2 H$

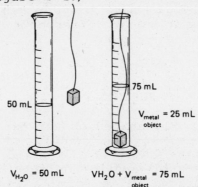

C-3 Calculate the volume in mm^3 and cm^3.   Show calculations
including units.

D. <u>VOLUME BY DISPLACEMENT</u>

An object placed under water will cause the water level to rise.
The difference in the water level before and after the object is
submerged is due to the volume of the object. The object has
displaced its own volume of water. By measuring the water level
before and after the object is added, the volume of the object can
be calculated. (See Figure 3-2.)

Figure 3-2   Volume of a Solid by Volume Displacement

D-1 Obtain a graduated cylinder that will fit the solid you used
in Part C. Partially fill the graduated cylinder with
water. Record this level as the initial volume of water.

D-2 Carefully place the solid used previously(Part C) in the
cylinder (Make sure the solid is completely covered by the
water.) Heavy solids should be tied to a piece of thread
and carefully lowered to the botton of the cylinder to
prevent cracking the glass. (If a solid floats, push it
under the surface of the water until it is just submerged.)
Record the <u>final level</u> of the water.

D-3 Calculate the volume for the solid displaced by subtracting the
initial level of the water from the final level.

D-4 State the volume of the solid in cubic centimeters.($1 \text{ mL} = 1 \text{ cm}^3$)

D-5 How does the volume obtained by direct measurement in Part
C-3 compare to the volume obtained by volume displacement in Part
D-4? Explain why might they might differ somewhat.

EXPERIMENT 3   THE METRIC SYSTEM: MEASURING VOLUME

LABORATORY RECORD

NAME_____
DATE_____
SECTION_____

PART I: READING THE VOLUME OF A FLUID IN A GRADUATED CYLINDER

A. <u>VOLUME</u> <u>OF</u> <u>LIQUIDS</u>

Volume of liquid (1)   _____ mL   _____ L

Calculation: _____mL   x $\dfrac{1\ L}{1000\ mL}$ =   _____ L

Volume of liquid(2)   _____ mL   _____L
Calculation:

Volume of liquid(3)   _____ mL   _____ L
Calculation:

B. <u>QUARTS</u> <u>AND</u> <u>LITERS</u>

B-1   Number of mL (measured) in 1 quart   _____ mL

B-2   Conversion Factor(mL/qt) $\dfrac{\text{_____ mL}}{1\quad qt}$

B-3   How does your experimental conversion factor for mL/qt compare
       with the given conversion factor, 946 mL/qt?

       Why might they be somewhat different?

EXPERIMENT 3   THE METRIC SYSTEM: MEASURING VOLUME

PART II: **VOLUME OF A SOLID**   NAME_____

C.   VOLUME BY DIRECT MEASUREMENT

C-1   Shape of Solid   _____

C-2   Dimensions: mm   cm
        (List type of
         measurement)

        _____   _____   _____

        _____   _____   _____

        _____   _____   _____

C-3   Volume of Solid: _____mm^3 _____cm^3

        Calculations:

D.   VOLUME BY DISPLACEMENT

D-1   Initial level of water   _____

D-2   Final level of water   _____

D-3   Volume of solid(mL)   _____

D-4   Volume of solid (cm^3)   _____
        Calculation:

              _____mL   x $\dfrac{1\ cm^3}{1\ mL}$ =   _____cm^3

D-5   How does the volume obtained by direct measurement in Part C-3
      compare to the volume obtained by volume displacement in Part
      D-4? Why might they differ somewhat?

1. Describe the method you would use to determine the volume of a
    solid with an irregular shape?

2. Using conversion factors, set up and solve the following
    problems. Use proper units and show unit cancellation.

    a. A patient has received 825 mL of fluid in one day. What is
        that volume in liters?

    b. A volume of 245 mL of a cleaning solution is mixed with
       0.575 L of water. What is the final volume in mL of the
       solution?

    c. How many quarts are in 5520 mL of sphagetti sauce?

    d. How many pints of plasma are present in 8.25 L? (1 qt = 2 pt)

    e. The water level in a graduated cylinder is 30.0 mL. A glass
       marble with a volume of 11.8 mL is added to the water. What
       is the final volume of water in the cylinder?

EXPERIMENT 4   THE METRIC SYSTEM: MEASURING MASS

PURPOSE

1. Use a laboratory balance to determine the mass of an object or substance.
2. Use the gram as a unit of mass.
3. Determine the conversion factors for grams and pounds; kilograms and pounds.

MATERIALS

laboratory balance
objects to weigh
 (stopper, metal piece, pencil, unknown, etc.)
small beaker
commerical product that has a label giving weight in U.S. customary
    units and mass in metric units
small (25 mL) graduated cylinder

KEYED OBJECTIVES IN TEXT: 1-1, 1-2, 1-3, 1-4, 1-5

DISCUSSION OF EXPERIMENT

The mass of an object indicates the amount of matter present in that object. The weight of an object is a measure of the attraction the earth has for that object. Since this attraction is proportional to the mass of the object, we will use the terms mass and weight interchangeably.

The unit of mass in the metric system is the gram (g). A larger unit, the kilogram (kg), may be used to measure a patient's weight in a hospital, while a smaller unit of mass, the milligram (mg) is often used in the laboratory. (See Table 4-1.)

Table 4-1   Some Metric Units used to Measure Mass

kilogram (kg)	1 kg = 1000 g	
gram (g)	1 g	
milligram (mg)	0.001 g	1 g = 1000 mg

LABORATORY ACTIVITIES

The Laboratory Balance

Your instructor will demonstrate the method of operation of a laboratory balance to be used in the laboratory. This may be a triple beam balance consisting of a weighing pan and a group of weights each resting on its own beam. You can then follow these general steps for weighing an object on a triple beam balance in Appendix A. If you have an electrical top-loading balance for laboratory use, see Appendix B for weighing instructions.

EXPERIMENT 4   THE METRIC: MEASURING MASS

PART I: FINDING THE MASS OF A SOLID OR LIQUID

A. THE LABORATORY BALANCE

Observe the laboratory balance you will be using in the lab.
Answer the questions for Part A in the laboratory record.

B. THE MASS OF A SOLID

B-1   Determine the mass of several solid objects. Where there is
no specific object listed, you may select one of your own
choice. Record each mass in grams(include units). Be sure
that all measurements include a figure in the hundredth's
place.

     Example: beaker   42.18 g   recorded to the
                                                hundredth's place

Be sure to put a 0 in the hundredths place when it is part of
the accuracy of the measurement.

     Example: pencil   11.60 g   recorded to the
                                                hundredth's place

B-2   When you have used the balance several times, ask your
instructor for the unknown object. This might be a piece of
metal with a code number on it. If so, record that code
number. Determine the mass of the unknown and check your
answer with the instructor.

C. MASS OF A LIQUID

The mass of a liquid can be determined by first weighing a beaker
and then weighing that beaker with the liquid. The difference in
the two weighings will give the mass of the liquid. This procedure
is called weighing by difference.

C-1   Weigh a small beaker. Record its mass.

C-2   Using a graduated cylinder, carefully measure out 25.0 mL
of water. Add the water to the beaker. (Do not use the
volume markings found on the sides of some beakers. These
are not for accurate volume measurement.) Weigh and record
the combined mass of the beaker and the water.

C-3   Calculate the mass of the water by the difference in the
two weighings. Record the mass of the water in grams.

C-4   Give a reason for using the weighing by difference method.

EXPERIMENT 4   THE METRIC SYSTEM: MEASURING MASS

PART II.   CONVERSION FACTORS FOR MASS

D. GRAMS AND POUNDS

D-1 Obtain a container of a commerical product whose label lists the mass of the product in both metric and U.S. customary units. Record. (Do not weigh. The mass refers to the substance that is/was in the container.)

D-2 If necessary, convert the metric unit to grams.

D-3 If necessary, convert the U.S. customary unit to pounds. (1 lb = 16 oz)

D-4 Calculate an experimental conversion factor for grams/pound (g/lb). Include both units in your final answers.

$$\frac{\text{number of grams}}{\text{number of lb}} = \underline{\hspace{2cm}} g/lb$$

D-5 State the given (standard) value of g/lb.

E. POUNDS AND KILOGRAMS

E-1 Using the same product label information from Part D, convert the metric unit to kilograms.

E-2 State the U.S. customary value in pounds.

E-3 Calculate an experimental conversion factor for pounds/kilogram (lb/kg).

$$\frac{\text{number of lb}}{\text{number of kg}} = \underline{\hspace{2cm}} lb/kg$$

E-4 State the given(standard) value of lb/kg.

PART III: LAB CONTEST - PERCENT DIFFERENCE (OPTIONAL)

Your instructor will display an object and everyone in the class will give an estimation of its mass. You may pick up the object, but you are not to weigh it. Record your estimation. When the object is weighed , record its actual value. How far off was your estimation? Determine the percent difference between your estimated value and the actual value.

$$\text{Percent Difference} = \frac{\text{Difference}}{\text{Actual Value}} \times 100$$

27

EXPERIMENT 4   THE METRIC SYSTEM: MEASURING MASS

LABORATORY RECORD

NAME_____
DATE_____
SECTION_____

PART I: FINDING MASS OF A SOLID OR A LIQUID

A. THE LABORATORY BALANCE

How would you "zero" a balance?

Where do you put the object to be weighed?

How large a mass can be measured by the balance you are using?

B. MASS OF A SOLID

B-1	**Object**	**Mass (grams)**
	one quarter	_____
	stopper	_____
	beaker	_____
	crucible	_____
	_____	_____
B-2	unknown(# _____)	_____
	checked by instructor	_____

C. MASS OF A LIQUID

C-1   mass of beaker   _____

C-2   mass of beaker and 25.0 mL water   _____

C-3   mass of 25.0 mL of water   _____

C-4   Give a reason for using the weighing by difference
       method.

29

EXPERIMENT 4   THE METRIC SYSTEM: MEASURING MASS
LABORATORY RECORD   NAME_____

PART II: CONVERSION FACTORS FOR MASS

D. GRAMS AND POUNDS

D-1   Commerical Product _____

        Mass on label: metric units_____ U.S. units_____

D-2   Metric unit(label) in grams(g)   _____

D-3 U.S Customary unit(label)in pounds(lb)   _____

D-4   Calculated conversion factor(g/lb)

            _____number of grams__ =   _____ g/lb
            _____number of lb

D-5   Standard conversion factor   _____g/lb

E. POUNDS AND KILOGRAMS

E-1   Metric unit(label) in kg   _____

E-2   U.S. Customary unit(label) in lb   _____

E-3   Calculated conversion factor (lb/kg)

            _____number of lb___   =   _____lb/kg
            _____number of kg

E-4   Standard conversion factor   _____lb/kg

PART III: LAB CONTEST - PERCENT DIFFERENCE (OPTIONAL)

        Estimated Mass of Object   _____

        Actual Mass of Object   _____

        Difference between estimated
        mass and actual mass   _____

        Percent Difference   _____
        Calculation:

30

QUESTIONS AND PROBLEMS

1.   What is the mass of an object when the balance reads 100.00 g,
     20.00 g, 8.00 g, 0.20 g and 0.05 g?

2.   What is the total mass of objects that have masses of
     0.200 kg, 80.0 g, and 524 mg?

3. A beaker has a mass of 225.08 g. When a liquid is added to the
   beaker, the combined mass is 278.25 g. What is the mass of the
   liquid alone?

4.Solve the following problems using conversion factors in a
   problem set up to cancel units.

   a.   How many mg is 0.078 g?

   b.   An infant has a mass of 3.40 kg. What is the weight of
        the infant in pounds?

4.   An adult has a mass of 58,500 g.

     a. What is that person's mass in kg?

     b. What is that person's weight in lb?

5.   Medication Problems

     a.   The doctor's order is Antabuse 240 mg. The stock on hand
          is 0.060-g tablets. How many tablets are needed to fill
          the order?

     b.   In the apothecary system, there are 60 mg in a grain (gr)
          or 60 mg/gr. The order is Mesurin 20 gr. The stock on hand
          is 600-mg tablets. How many tablets are to be administered?

     c. The order is Pen V K 500 mg. The label reads 250 mg per
        teaspoon (250 mg/tsp); 1 tsp equal 5 mL (5 cc). How many mL
        of the medicine should be given?

EXPERIMENT 5   DENSITY AND SPECIFIC GRAVITY

PURPOSE

1. Calculate the density of a solid or liquid by measuring the mass and volume.
2. Calculate the specific gravity of a liquid.
3. Check the specific gravity of a liquid using a hydrometer.
4. Predict from calculated densities whether objects will sink or float in different liquids.
5. Design an experiment to collect data for the preparation of a graph to illustrate the relationship between mass and volume of a substance.

MATERIALS

50-mL graduated cylinder
100-mL beaker
isopropyl alcohol
water
ice cubes

Solid objects of wood and metal
hydrometers in liquids
Substance for design experiment:
     lead shot, unpopped popcorn,
     glass marbles, nails, beans,
     etc.

KEYED OBJECTIVES IN THE TEXT: 1-6

DISCUSSION OF EXPERIMENT

The density of a substance represents a relationship between the mass of a substance and its volume. Hospital laboratories determine the density of urine as part of a health checkup. The fluid in your car battery is checked using a hydrometer to evaluate the condition of the battery. The observation that a substance sinks in one liquid and floats in another can be explained by comparing the densities of the substances.

To determine the density of any substance, both the mass and the volume must be measured. You have carried out both of these procedures in previous experiments. When the mass and volume are known, the density can be calculated by using the following relationship:

$$\text{Density} = \frac{\text{mass}}{\text{volume}}$$

If the mass is measured in grams, and the volume measured in milliliters, the density will have the units of g/mL.

EXPERIMENT 5   DENSITY AND SPECIFIC GRAVITY

LABORATORY ACTIVITIES

PART I: **DENSITY**

A. DENSITY OF SOLIDS

A-1 Obtain a wood block and a piece of metal. Weigh each on the
     laboratory balance. Record the mass of each. (Remember that
     your measurements must have two figures after the decimal
     points as well as units.)

A-2 Determine the volume of each object by volume displacement.
     This procedure is explained in experiment 3.

A-3 Calculate the density of each solid by dividing the mass by
     the volume. The units are g/mL. Record each value.

$$\text{Density} \quad = \quad \frac{\text{mass in grams}}{\text{volume in mL}}$$

B. DENSITY OF LIQUIDS

Determine the mass, volume and density of water and isopropyl
alcohol.

B-1 Weigh a 100-mL beaker and record its mass.

B-2 Using a graduated cylinder, add a carefuly measured volume
     such as 20.0 ml of the liquid to the beaker and weigh. Record
     the combined mass.

B-3 Calculate the mass of liquid.

B-4 Record the volume of liquid used.

B-5 Calculate the density of the liquid by dividing the mass by
     the volume of the liquid.

$$\text{Density} \quad = \quad \frac{\text{mass}}{\text{volume}}$$

PART II: SPECIFIC GRAVITY

The specific gravity of a liquid compares its density to the density
of water.

$$\text{specific gravity (sp gr)} \quad = \quad \frac{\text{density of fluid (g/mL)}}{\text{density of water (g/mL)}}$$

This comparison results in a number with no units. This is one of
the few measurements in chemistry that has no units; the units of
density have cancelled out.

EXPERIMENT   DENSITY AND SPECIFIC GRAVITY

Using a Hydrometer

Specific gravity can also be measured with a hydrometer. Small hydrometers (urinometers) may be used in the hospital to determine the specific gravity of urine. Another type of hydrometer is used to measure the fluid level in your car battery.

A hydrometer is set in a liquid and spun slowly to keep it from sticking to the sides of the container. The specific gravity scale is read at the meniscus of the fluid. Read the hydrometer scale to the one-thousandths place which is three figures after the decimal point. (See Figure 5-1.) (Some hydrometer may use the European decimal point which is a comma. Record this as a decimal point.)

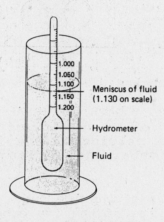

Figure 5-1.   Reading a hydrometer.

C.   CALCULATING SPECIFIC GRAVITY

C-1   Calculate the specific gravity (sp gr) for water and isopropyl alcohol from the density you calculated in Part B.   Divide each density by the standard value of H_2O at 4°C which is 1.00 g/mL.

$$\frac{D_{liquid}(g/mL)}{D_{H_2O} (1.00 \text{ g/mL})} = \text{sp gr}$$

D.   SPECIFIC GRAVITY USING A HYDROMETER

D-1   Read the hydrometers set in graduated cylinders of the same liquids.   Record your hydrometer values for the specific gravities of water and isopropyl alcohol.

D-2   Compare the speicific gravities of water and isopropyl alcohol that you calculated in part C with the hydrometer readings for their specific gravities. Explain why they might differ.

35

EXPERIMENT 5   DENSITY AND SPECIFIC GRAVITY

PART III: SINK OR FLOAT

E. PREDICTIONS
    Let us consider the hypothesis that if an object is less dense
    than a liquid, then that object will float on the liquid.
    Using this hypothesis and the densities calculated for the
    solid and the liquids, predict whether the wood and the piece
    of metal from Part A will sink(S) or float(F) in each of the
    liquids in Part B.

F. CHECKING YOUR PREDICTIONS
    This may be a demonstration by your instructor at the end of
    the lab, or you may wish to use the materials found on the
    demonstration table. Each of the solids has been tied by a
    thread. Hold one of the samples by the thread and slowly
    lower the object into a beaker containing one of the liquids.
    Record the actual result as sink (S) or float (F) next to your
    prediction. Continue testing the solid with the other
    liquids. Repeat the testing with another one of the solids
    until you have checked all of your predictions.

G. WHY DOES AN ICE CUBE FLOAT? (OPTIONAL)
    The determination of the density of an ice cube is complicated
    by the continual melting of the ice cube. However, good
    results can be obtained by planning your steps and then
    measuring quickly.

G-1 Determine the mass of the ice cube.
G-2 Determine the volume of the ice cube by volume displacement.
G-3 Calculate the density of the ice cube.
G-4 Predict whether the ice cube should sink or float in each
     of the liquids.
G-5 Check your predictions.

H. DESIGN AN EXPERIMENT(OPTIONAL)
Design an experiment to obtain the mass, volume and density of four
samples of the same substance such as lead shot, unpopped popcorn,
glass marble, nails, or some other substance of your choice. Then
graph the data to illustrate the relationship between mass and
volume. To assist you in your design, consider the following steps.

A. State the purpose of the experiment.
B. Describe the experimental procedure including steps for the
    measurement of the mass and volume and calculations of the
    density of each sample.
C. Design data tables for measurements and calculations.
D. Graph the mass(g) of each sample on the vertical axis against
    the measured volume(ml) on the horizontal axis. See Appendix C
    for guidelines to graphing. There is graph paper at the back of
    this lab book.
E. State conclusions for your experiment and graph. Include some
    evaluation of sources of error.

EXPERIMENT 5   DENSITY AND SPECIFIC GRAVITY

LABORATORY RECORD   NAME_____
                                           DATE_____
PART I: DENSITY   SECTION_____

A. DENSITY OF SOLIDS

                             Wood   Metal
                             Block   Piece

A-1   Mass (g)   _____   _____

A-2   Final water level _____   _____

       Initial water level _____   _____

       Volume of object(mL) _____   _____

A-3 Density (g/mL)   _____   _____

                            Calculations:

B.   DENSITY OF LIQUIDS
                             Water   Alcohol

B-1 Mass of beaker   _____   _____

B-2 Mass of beaker and liquid _____   _____

B-3 Mass of liquid   _____   _____

B-4 Volume of liquid   _____   _____

B-5 Density(g/mL)   _____   _____

                            Calculations:

PART II: SPECIFIC GRAVITY

LIQUIDS

Water   Alcohol

C. Specific Gravity
    (Calculated)   _____   _____

    Calculations:

D.   Specific Gravity
      using hydrometers   _____   _____

How do your calculated specific gravities compare with the
specific gravity readings you obtained by using hydrometers?

Why might the calculated values be somewhat different than the
values obtained using hydrometers?

PART III: SINK OR FLOAT

E., F.   SOLIDS
                          Wood   Metal
                          Block   Piece

LIQUIDS	PREDICTED	ACTUAL	PREDICTED	ACTUAL
Water	_____	_____	_____	_____
Alcohol	_____	_____	_____	_____

EXPERIMENT 5 DENSITY AND SPECIFIC GRAVITY   NAME_____

G. <u>WHY</u> <u>DOES</u> <u>AN</u> <u>ICE</u> <u>CUBE</u> <u>FLOAT?</u>(OPTIONAL)

    G-1   Mass of the ice cube   _____

    G-2   Volume of the ice cube   _____
               Data and calculations for volume determination:

    G-3   Density of the ice cube   _____
               Calculations:

    G-4, G-5   Ice cube

	PREDICTED	ACTUAL
Water	_____	_____
Alcohol	_____	_____

Does an ice cube sink or float in water? Why?

Does an ice cube sink or float in alcohol? Why?

H. DESIGN AN EXPERIMENT

A. PURPOSE

B. EXPERIMENTAL PROCEDURE

C. DATA TABLES

D. GRAPH DATA ON STANDARD GRAPH PAPER FOUND AT BACK OF THIS LAB
    BOOK

E. CONCLUSIONS AND SOURCES OF ERROR

EXPERIMENT 5   DENSITY AND SPECIFIC GRAVITY   NAME_____

QUESTIONS AND PROBLEMS

1. An object made of aluminum has a mass of 8.37 g. When it was placed in a graduated cylinder containing 20.0 mL of water, the water level rose to 23.1 mL. Calculate the density of the object.

2. What is the mass of a solution that has a density of 0.775 g/mL and a volume of 50.0 mL?

3. A salt solution has a mass of 80.25 g and a volume of 75.0 mL. What is the specific gravity of that solution?

4. A piece of plastic has a density of 1.25 g/mL. Using density comparison, indicated whether you expect the piece of plastic to sink or float in the following liquids. (DO NOT TEST YOUR PREDICTIONS):

    Water(D= 1.00 g/mL) _____   Ethyl alcohol(D= 0.79 g/mL)_____

    Carbon tetrachloride(D=1.60 g/mL)_____

5. You have a liquid in a small container, a plastic cube, a balance,   and several graduated cylinders. The plastic cube is too large to fit into the container holding the liquid or in any of the graduated cylinders. Describe the measurements and calculations that you could use to determine if the plastic cube will sink or float in the liquid.

EXPERIMENT 6 TEMPERATURE MEASUREMENT

PURPOSE

1. Operate a Bunsen burner properly.
2. Set up a water bath correctly.
3. Measure temperature with a thermometer using the Celsius temperature scale.
4. Convert temperature readings to their corresponding Fahrenheit and Kelvin temperatures.
5. Graph a change in temperature against units of time.

MATERIALS

Bunsen burner	wire screen
thermometer	ringstand and iron ring
250-ml beaker	striker or match
ice	timer (or watch with second hand)

KEYED OBJECTIVE IN TEXT: 1-7

DISCUSSION OF EXPERIMENT

Temperature is a measure of the intensity of heat in a substance. A substance with little heat feels cold. Where the heat intensity is great, a substance feels hot. The temperature of our bodies is an indication of the heat produced from normal metabolism. A condition such as an infection may cause body temperature to deviate from normal.

In the laboratory, temperature is usually measured by the Celsius scale. The Celsius temperature can be converted to its corresponding temperature on the Fahrenheit scale by using the following equation:

$$T^o_F = 1.8 \, (T^o_C) + 32$$

The Celsius temperature may also be converted to its temperature on the metric scale which is the Kelvin scale. The units on this scale are called kelvins.

$$T_K = T^o_C + 273$$

43

EXPERIMENT 6   TEMPERATURE MEASUREMENT

LABORATORY ACTIVITIES

PART I: MEASURING TEMPERATURE

A. THE LABORATORY BURNER

In chemistry, heat is usually supplied by the Bunsen burner. While
there are several models, they are similar in design and function.
The typical burner consists of a tube or barrel through which a
fast-moving stream of gas (natural gas) and air flows. At the
bottom of the barrel is the air intake. Closing and opening the
air vent regulates the amount of air mixing with the gas. The
burner is connected to the gas source at the desk by a piece of
rubber tubing. The amount of gas is controlled by a wheel or screw
at the base of the burner which is attached to a needle valve.
Turning this wheel or screw into the burner limits the flow of
gas; turning it out increases the flow of gas.   Gas flow may also
be regulated at the desk source by opening or closing the gas
valve. Make sure the gas valve is tightly closed when you leave
the laboratory after using the Bunsen burner. See Figure 6-1.

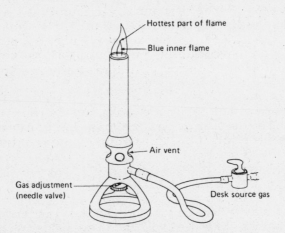

Figure 6-1   Laboratory Burner

Before you begin lighting the burner, practice (a) opening and
closing the gas cock at the desk source of gas; (b) opening and
closing the gas valve(s); and (c) opening and closing the air vent.

**SAFETY NOTE: A Bunsen burner is a potential hazard. Be sure that
long hair is tied back. Keep the work area clear of books, papers,
and other flammable items. Always keep the burner on the desk top.
Never stack books up to hold the burner at a higher level.**

PROCEDURE:

A-1 Close the air intake by rotating the barrel of the burner or
    closing the holes of the air vent. Prepare to light the burner
    by having a striker or match ready. Your instructor may

44

demonstrate the use of the striker. Turn the gas on and hold
the flame or spark at the top of the burner. If you are using a
striker, be sure you strike the flint hard enough to get a
spark. Practice lighting the burner a few times.   Adjust the
gas supply to alter the height of the flame. Answer the
questions in the laboratory record.

A-2 Slowly open the air intake. The flame should change from a
    yellow, sooty flame to a bluish flame. For general heating,
    you obtain the most heat from a flame when you adjust the air
    vent to give a flame that has an inner blue section. Right
    above this inner blue section is the hottest part of the flame
    and this is what you should use for heating. Opening the air
    vents too much will make the burner noisy or will blow out the
    flame. Answer the questions in the laboratory record.

B.   MEASURING TEMPERATURE

The laboratory thermometer is larger than the thermometer you use
at home or in the hospital.   Laboratory thermometers respond
quickly to their surroundings. When you are measuring the
temperature of a solution or substance, always read the thermometer
while it is **immersed** in that solution or substance.

**SAFETY NOTE: NEVER SHAKE A LABORATORY THERMOMETER. There is no
need to shake a laboratory thermometer because it contains no
restrictions to the flow of mercury as do the oral thermometers.
Shaking a laboratory thermometer can cause serious accidents.**

B-1 Obtain a thermometer. Observe the markings on the thermometer.
     Answer the questions on the record sheet.

B-2 Using the Celsius scale, measure the temperature of the
     following:
     a.   Tap water placed in a beaker.
     b.   An ice slurry (crushed ice and water).
     c.   A salted ice slurry (Add rock salt to ice slurry).
     d.   Room temperature (place the thermometer on the lab bench).
     e.   Boiling water (this may be done when you boil water in the
          next section of the experiment)

     Complete the temperature table by converting the Celsius
     temperatures to their corresponding temperatures on the
     Fahrenheit and Kelvin scales.

PART II: GRAPHING A CHANGE IN TEMPERATURE WITH TIME

Graphing is an important skill for describing information in a
visual manner. In the hospital, graphs are used to plot the growth
of children, to monitor the effect of medication, or to follow
fluctuations in body temperature. Drawing and reading graphs
properly is an important skill for anyone working in the health
fields.   See Appendix C for graphing instructions.

EXPERIMENT 6   TEMPERATURE MEASUREMENT

C.   COLLECTING MEASUREMENTS FOR THE DATA TABLE

Set up a water bath using an iron ring, a wire gauze, a 250-mL
beaker containing 150-mL of water. The height of the iron ring
should be about 2 inches above the burner. (The tip of the inner
flame should touch the bottom of the beaker.)   Do not rest the
thermometer on the side or bottom of the beaker. Support the
thermometer by tying a string to the loop in the top of the
thermometer or using a clamp securely fastened to the thermometer.
To take temperature measurements, the bulb of the thermometer
should be centered in the liquid.   See Figure 6-2.

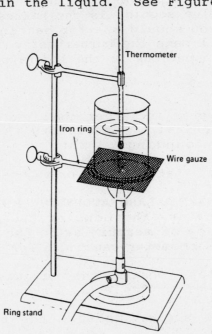

Figure 6-2   Apparatus for a boiling water bath.

     Using a watch with a second hand or a timer obtained from the
stockroom, record the temperature of the water every 0.5 min (30
sec). Your first measurement occurs at 0 min with the correct
temperature of the water. Begin heating the water.   Continue
temperature measurements and timing until the water has been at a
full boil with a constant temperature for at least 2 min. (Record
the boiling temperature of water on the temperature table B-2.

D.   GRAPHING THE CHANGE IN TEMPERATURE

Read the directions in Appendix C for the construction of a graph.
Using your data from the heating table, construct a graph with
equal intervals of temperature($^{\circ}$C) on the vertical axis, and
time(min) on the horizontal axis.   Graph paper can be found at the
back of this lab book.

LABORATORY RECORD

PART I: MEASURING TEMPERATURE

A. THE LABORATORY BURNER

A-1 How do you control the height of the flame?

A-2 What happens to the color of the flame as the air vent is opened?

What type of flame should you use for general heating with the burner?

B. READING THE THERMOMETER

B-1 What temperature scale(s) are represented on the thermometer?

What are the highest and the lowest temperatures that can be measured using your thermometer?

B-2 Temperature Table

| | TEMPERATURE | | |
	°C	°F	K
(a) tap water	_____	_____	_____
(b) ice slurry	_____	_____	_____
(c) salt-ice slurry	_____	_____	_____
(d) room	_____	_____	_____
(e) boiling water	_____	_____	_____

47

PART II: GRAPHING A CHANGE IN TEMPERATURE WITH TIME

C.   DATA TABLE

Time (min)	Temperature °C	Time(min)	Temperature °C
0.0	_____	10.5	_____
0.5	_____	11.0	_____
1.0	_____	11.5	_____
1.5	_____	12.0	_____
2.0	_____	12.5	_____
2.5	_____	13.0	_____
3.0	_____	13.5	_____
3.5	_____	14.0	_____
4.0	_____	14.5	_____
4.5	_____	15.0	_____
5.0	_____	15.5	_____
5.5	_____	16.0	_____
6.0	_____	16.5	_____
6.5	_____	17.0	_____
7.0	_____	17.5	_____
7.5	_____	18.0	_____
8.0	_____	18.5	_____
8.5	_____	19.0	_____
9.0	_____	19.5	_____
9.5	_____	20.0	_____
10.0	_____		

QUESTIONS AND PROBLEMS

1. Your friend has a temperature of 312.5 K. What is your
    friend's temperature in degrees Fahrenheit?   (Show work)

2. Some laboratory experiments are run at -100°F. What is that
    temperature in degrees Celsius?

3. Use the graph you constructed in Part II of this experiment to
    answer the following:

    a. How many minutes were needed for the temperature to rise to
        60°C?

    b. How many minutes did it take for the temperature to change
        from 50°C to 75°C?

    c. Eventually, during the heating, the temperature reaches a
        constant value and no longer increases. What physical change
        was taking place in the beaker?

    d. What would be the name of the temperature at which boiling
        occurs?

PURPOSE

1. Write the correct symbols or names of some elements.
2. Describe the physical properties you observe for some elements.
3. Categorize an element as a metal or nonmetal using
   (a) physical properties and (b) location on the periodic table.

MATERIALS

Display of elements

KEYED OBJECTIVES IN TEXT   2-1,2-2

DISCUSSION OF EXPERIMENT

Primary substances, called <u>elements</u>, build all the materials about
you. There are 109 elements known today. The smallest unit
characteristic of any element is an <u>atom</u> of that element. There is
a different kind of atom for every element.

Take a look at some elements in the display. Some look
very different from each other, but others look similar. Elements
can be categorized in several ways. In this experiment, you are to
group elements by similarities in their physical properties.
Elements that appear shiny or lustrous are called <u>metals</u>. They are
usually good conductors of heat and electricity, somewhat soft and
ductile, and can be molded into a shape. Some of the metals you
will examine such as sodium or calcium may have an outer coating of
a white oxide formed by combination with oxygen in the air. If
cut, you could see the fresh shiny metal underneath. Other
elements are not good conductors of heat and electricity, are
brittle, and appear dull, not shiny. These element are called
<u>nonmetals</u>.

LABORATORY ACTIVITIES

PART I: ELEMENTS

A. <u>SYMBOLS</u> <u>AND</u> <u>NAMES</u>

Complete the chart by listing the symbol and atomic number for
each element. Using the display of elements and the list in
the record sheet, describe the color, luster and any other
features you observe for the elements.

EXPERIMENT 7   ELEMENTS AND THEIR SYMBOLS

PART II: METALS AND NONMETALS

B.   PERIODIC TABLE

On the periodic chart provided in the laboratory record, draw a circle around the symbols of those elements listed in Part I that are shiny. Draw a square around the symbols of the elements from the list (Part I) that are dull (not shiny), or are gases.

C.   METALS AND NONMETALS

The circles you drew on the periodic chart indicate elements that are metals. The squares are usually drawn around the symbols of elements that are nonmetals. This distinction is often indicated by a heavy zig-zag line which looks like a staircase on the periodic chart which begins to the left of boron (B).

C-1 The terms metals and nonmetals are not really that definite, but the heavy zig-zag line helps to separate those elements with typically more metallic behavior from those elements that show more nonmetallic behavior. However, you may find some overlapping of circles and squares around this heavy line. Use this information to answer the question in the laboratory record.

C-2 DO NOT LOOK AT THE DISPLAY OF ELEMENTS. By the location on the periodic table, predict whether the elements listed would be (a) metals or nonmetals and (b) shiny or dull. Now, observe those elements in the display to see if your predictions are correct.

EXPERIMENT 7: ELEMENTS AND SYMBOLS: METALS AND NONMETALS

LABORATORY RECORD

NAME _____

DATE _____

PART I: ELEMENTS

SECTION _____

A. Element	Symbol	Atomic Number	B. Physical Properties(color,luster,other)
zinc	_____	_____	_____
copper	_____	_____	_____
carbon	_____	_____	_____
oxygen	_____	_____	_____
aluminum	_____	_____	_____
phosphorus	_____	_____	_____
silver	_____	_____	_____
iron	_____	_____	_____
sulfur	_____	_____	_____
magnesium	_____	_____	_____
silicon	_____	_____	_____
tin	_____	_____	_____

PART II: METALS AND NONMETALS

B. <u>PERIODIC TABLE</u>

Atomic number, Symbol, Atomic Mass

1
H
1.0

Group IA	IIA											IIIA	IVA	VA	VIA	VIIA	VIIIA
																	2 He 4.0
3 Li 6.9	4 Be 9.0											5 B 10.8	6 C 12.0	7 N 14.0	8 O 16.0	9 F 19.0	10 Ne 20.2
11 Na 23.0	12 Mg 24.3				Transition elements (B)							13 Al 27.0	14 Si 28.1	15 P 31.0	16 S 32.1	17 Cl 35.5	18 Ar 39.9
19 K 39.1	20 Ca 40.1	21 Sc 45.0	22 Ti 47.9	23 V 50.9	24 Cr 52.0	25 Mn 54.9	26 Fe 55.8	27 Co 58.9	28 Ni 58.7	29 Cu 63.5	30 Zn 65.4	31 Ga 69.7	32 Ge 72.6	33 As 74.9	34 Se 79.0	35 Br 79.9	36 Kr 83.8
37 Rb 85.6	38 Sr 87.6	39 Y 88.9	40 Zr 91.2	41 Nb 92.9	42 Mo 95.9	43 Tc (99)	44 Ru 101.1	45 Rh 102.9	46 Pd 106.4	47 Ag 107.9	48 Cd 112.4	49 In 114.8	50 Sn 118.7	51 Sb 121.8	52 Te 127.6	53 I 126.9	54 Xe 131.3
55 Cs 132.9	56 Ba 137.3	57 La 138.9	72 Hf 178.5	73 Ta 180.9	74 W 183.8	75 Re 186.2	76 Os 190.2	77 Ir 192.2	78 Pt 195.1	79 Au 197.0	80 Hg 200.6	81 Tl 204.4	82 Pb 207.2	83 Bi 209.0	84 Po (210)	85 At (210)	86 Rn (222)
87 Fr (223)	88 Ra (226)	89 Ac (227)	104 Ku (257)	105 Ha	106	107		109									

Note numbers in parentheses indicate mass number of most stable or best known isotope

53

EXPERIMENT 7 ELEMENTS AND THEIR SYMBOLS

C. METALS AND NONMETALS

C-1 LOCATION ON THE PERIODIC TABLE

Observe the periodic chart you completed in part B and describe the location of the metals and nonmetals relative to the heavy zig-zag line.

C-2 PREDICTION OF APPEARANCE OF SOME OTHER ELEMENTS

| Element | _____ Predictions _____ | | |
	Metal or Nonmetal	Shiny or Dull	Correct?
chromium(Cr)	_____	_____	_____
gold (Au)	_____	_____	_____
lead(Pb)	_____	_____	_____
boron(B)	_____	_____	_____
bromine(Br)	_____	_____	_____

QUESTIONS AND PROBLEMS

1. Complete the list of names of elements and symbols:

Element	Symbol	Element	Symbol
potassium	_____	_____ Na
sulfur	_____	_____ P
nitrogen	_____	_____ Fe
magnesium	_____	_____ Cl
copper	_____	_____ Ag

2. What are some differences in the physical properties between metals and nonmetals?

3. What feature on the periodic table helps you to determine that an element is a metal or a nonmetal?

4. Use the periodic table to categorize each of the following elements as a metal or nonmetal.

Na _____   S _____

F _____   Fe _____

Ca _____   O _____

C_____   Zn_____

EXPERIMENT 8   ATOMIC STRUCTURE AND PERIODIC LAW

PURPOSE

1. Describe an atom in terms of atomic number, mass number, number of protons, neutrons, and electrons.
2. Draw a graph of atomic diameter against atomic number.
3. Interpret the trends in atomic diameter within a family and a period.

KEYED OBJECTIVES IN TEXT: 2-2, 2-3, 2-4, 2-5

DISCUSSION OF EXPERIMENT

Atoms are made of several subatomic particles. Three important subatomic particles are the proton, neutron and electron. Protons are positively charged, electrons are negatively charged, while neutrons are neutral (no charge). Inside the atom, the protons and neutrons are contained in a tightly packed ball called the **nucleus**. The rest of the atom which is mostly empty space is occupied by fast-moving electrons. Electrons are so small that their mass is considered to be negligible compared to the mass of the proton or neutron.

The number of protons and electrons determine the kind of atom for every element. The number of protons is equal to the **atomic number** of the element.

        ATOMIC NUMBER = NUMBER OF PROTONS

Protons attract electrons because electrons have an opposite charge. In a neutral atom, there are an equal number of protons and electrons.
        IN A NEUTRAL ATOM
    THE NUMBER OF PROTONS = THE NUMBER OF ELECTRONS

Atoms are also identified by their **mass number**. The **mass number** indicates the total number of protons and neutrons.

    MASS NUMBER   =   TOTAL NUMBER OF PROTONS AND NEUTRONS

Atoms can have some variation in the numbers of neutrons contained in the nucleus. All of the atoms of the same element will have the same number of protons, but they can differ in the number of neutrons. This means that atoms of the same element can have different atomic masses. They are called **isotopes**.

Since the 1800s, scientists have recognized that chemical and physical properties of certain groups of elements tend to be similar. A Russian scientist, Dmitri Mendeleev, found that the chemical properties of elements tended to occur periodically when the elements were arranged in order of increasing atomic mass. He used this periodic pattern to predict the behavior of elements that were not yet discovered. Later, H.G. Moseley established that the similarities in properties were associated with the atomic number.

57

LABORATORY ACTIVITIES

PART I: DETERMINING THE NUMBERS OF SUBATOMIC PARTICLES

Complete the table given in the laboratory record with the correct atomic numbers, mass numbers, number of protons, neutrons and electrons.

PART II: ISOTOPES

From the nuclear symbols of atoms of an element, state the number of protons, neutrons and electrons for each atom.

PART III: GRAPHING A PERIODIC PROPERTY

In this exercise, you will be graphing the relationship between the atomic diameter of an atom and its atomic number. Such a graph will show a repeating or periodic trend. In order to explain a graph of this type, observe the graph that is obtained by plotting the average temperature of the seasons. See Figure 8-1.

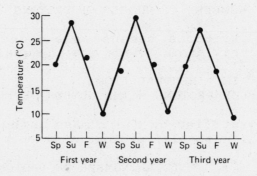

Figure 8-1   A graph of average seasonal temperatures.

The preceding graph indicates a cycle of temperature that is repeated each year. Such a tendency is known as a **periodic characteristic.** There are three cycles on this particular graph, one full cycle occuring every year. When such cycles are known, the average temperatures for the next year could be predicted.

EXPERIMENT 8   ATOMIC STRUCTURE AND PERIODIC LAW

GRAPHING THE ATOMIC DIAMETER AGAINST THE ATOMIC NUMBER

Atomic diameters can be plotted against atomic numbers to illustrate the concept of the periodic tendencies of the elements. The atomic diameters for the first 25 elements are given below. Using the data in Table 8-1, plot the atomic diameter against the atomic number of each element on the graph provided in the laboratory record. Graph paper can be found at the back of this lab book. Remember to place a title on your graph and to draw a smooth line. Use the graph you prepare to answer the questions in the laboratory record.

Use the graph you prepare to answer the questions in the laboratory record.

Table 8-1   Atomic Diameters for the First 25 Elements

Period	Element	Atomic Number	Atomic Diameter ($\overset{\circ}{A}$)*
1	H	1	0.7
	He	2	1.0
2	Li	3	3.0
	Be	4	2.2
	B	5	1.8
	C	6	1.5
	N	7	1.4
	O	8	1.3
	F	9	1.2
	Ne	10	1.4
3	Na	11	3.7
	Mg	12	3.2
	Al	13	2.9
	Si	14	2.3
	P	15	2.2
	S	16	2.1
	Cl	17	2.0
	Ar	18	1.9
4	K	19	4.6
	Ca	20	3.9
	Sc	21	3.2
	Ti	22	2.9
	V	23	2.7
	Cr	24	2.5
	Mn	25	2.5

*$\overset{\circ}{A}$ = 10^{-8} cm

EXPERIMENT 8 ATOMIC STRUCTURE

LABORATORY RECORD

NAME _____

DATE _____

SECTION _____

PART I: DETERMINING THE NUMBERS OF SUBATOMIC PARTICLES

Symbol	Atomic number	Mass number	Neutrons	Protons	Electrons
F		19			
Fe			30		
		27			13
K			20		
Br		80			
			74	53	

PART II: ISOTOPES

	Protons	Neutrons	Electrons
40 Ca 20			
42 Ca 20			
43 Ca 20			
44 Ca 20			
46 Ca 20			

QUESTIONS

1.   An atom of chlorine has a mass number of 37. Describe how you
     can determine the number of neutrons in the atom.

2. In Part II, you determined the number of subatomic particles for
     several atoms of calcium. Discuss why these atoms are
     called isotopes.

3. (a) What is the general change in atomic diameter as you go across
     the second period from Li(At. No. 3) to Ne(At. No.10)?

   (b) Would you expect the same change in the third period from Na
     to Ar?

     How is your prediction verified by your graph?

   (c) Why is this change called a periodic property?

EXPERIMENT 9   FLAME TESTS AND ELECTRON ARRANGEMENT

PURPOSE

1. Identify the color of a flame produced by an element.
2. Use the color of a flame to identify an element.
3. Draw a model of an atom including the electron arrangement for the first 20 elements.

MATERIALS

For flame testing:
  platinum wire
  Bunsen burner
  solutions (0.1 M)
    $CaCl_2$, KCl, $BaCl_2$, $SrCl_2$, $CuCl_2$, NaCl
    unknown solution

6M HCl(CAUTION: CORROSIVE)
small beaker
spot plate

For atom models: (optional)
  watch glass
  chalk
  items to represent neutrons, protons and electrons such as seeds, beans, paper clips, buttons, glass beads, etc.

KEYED OBJECTIVES IN TEXT: 2-6, 2-7

Certain elements produce strong colors in their flames when heated. Normally, there is a specific arrangement of the electrons in the energy levels occupied by the electrons. However, during heating one or more electrons may absorb energy in sufficient amounts (quanta) to jump to a higher (but less stable) energy level.

When an electron drops from a higher energy level to a lower energy level, a certain amount of energy is released. If that energy is emitted (released) as visible light, then a color is seen in the flame of the element. This color is very specific for an element and serves as a method for identifying elements. For other elements, energy may also be released by the change of energy level, but it may not be in the visible range. For example, energy in the form of heat, x-rays, infrared and ultraviolet light are all in the nonvisible energy spectrum.

The chemistry of an element depends on the arrangement of the electrons. The energy levels for electrons of atoms of the first 20 elements have the following arrangement:

Energy Level	Electrons
1	2e
2	8e
3	8e
4	2e

EXPERIMENT 9   FLAME TESTS AND ELECTRON ARRANGEMENT

PART I: FLAME TESTS

PREPARATION

Obtain the materials for flame tests. Bend the end of the flame-test wire into a small loop and secure the other end in a small cork stopper. Pour a small amount of 6 M (6 N) HCl into a beaker and clean the wire by dipping the loop in the HCl. Adjust the flame of a Bunser burner until it is colorless. Place the loop in the flame of the Bunsen burner. If you see a strong color in the flame while heating the wire, clean and heat the wire again until the color is gone.

Using a spot plate (a small plate with indentations), carefully place small amounts of the solutions $CaCl_2$, KCl, $BaCl_2$, $SrCl_2$, $CuCl_2$, and NaCl in different spots. Be careful not to mix the different solutions. Mark these on the spot plate diagram in the laboratory record.

HEATING

Dip the cleaned wire in the first solution on the spot plate. Make sure that a thin film of the solution adheres to the loop. (See Figure 9-1.) Move the loop of the wire into the lower portion of the flame and record the color you observe. If the color is not intense enough, use a clean dropper to place a drop of the solution on the loop, and test again. The color of the KCl flame is short-lived. Be sure to observe the color of the flame from the KCl solution in the first few seconds of heating. Clean the wire and repeat the flame tests with the other solutions.

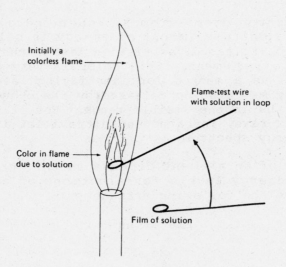

Figure 9-1   Flame test.

EXPERIMENT 9   FLAME TESTS AND ELECTRON ARRANGEMENT

TESTING AN UNKNOWN SOLUTION

Obtain an unknown solution that contains one of the above
solutions. Using the flame test procedure, determine the color
of the flame of your unknown. You may wish to compare the
flame color to the color of the flame tests for the known
solutions you looked at earlier. For example, if you think
your unknown is KCl, recheck the color of the KCl solution for
verification. Indicate the element that was responsible for
the color of the flame for your unknown.

PART II: DRAWING MODELS OF ATOMS

A model of an atom can be represented by a drawing that
indicates the nucleus and its protons and neutrons with the
electrons shown as concentric circles.   This is an
oversimplification of the true nature of atoms, but the model
does serve to illustrate the numerical relationship of the
subatomic particles. For example, a drawing of an atom of
boron with a mass number of 11 might look like this. Boron has
atomic number 5 which means it has 5 protons. The number of
neutrons is determined by substracting the number of protons
(5) from the mass number (11) to give 6 neutrons.

MODEL	NUCLEAR SYMBOL

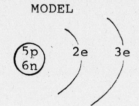

11   (mass number)
 B   (symbol of element)
 5   (number of protons)

To help you draw these models, you may wish to
construct visual examples of some of the atoms. A watch glass
can serve as a nucleus. The energy levels can be drawn with
chalk on the top of your desk or on paper covering the desk.
Obtain materials such as dried beans or peas, seeds or other
small objects. They can represent neutrons, protons and
electrons. Set up a model on the desk using these items to
portray an atom you wish to draw.

Draw a model of each atom listed on the laboratory record. The
nuclear symbol for each atom is given.

EXPERIMENT 9   FLAME TESTS AND ELECTRON ARRANGEMENT

LABORATORY RECORD

NAME_____
DATE_____
SECTION_____

PART I: FLAME TESTS

spot plate

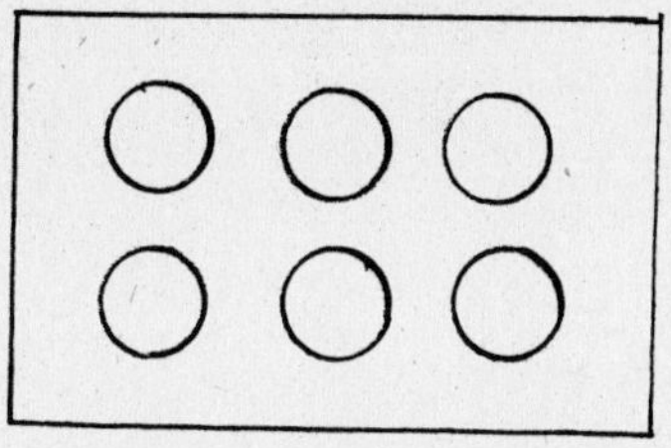

Solution	Flame Test Color
$CaCl_2$	_____
KCl	_____
$BaCl_2$	_____
$SrCl_2$	_____
$CuCl_2$	_____
NaCl	_____

Unknown Solution (code number _____)

Flame Color _____

Identification of Unknown _____

67

PART II: DRAWING MODELS OF ATOMS   NAME_____

7 Li 3	14 N 7	20 Ne 10
19 F 9	24 Mg 12	1 H 1
24 Na 11	31 P 15	37 Cl 17

EXPERIMENT 9   FLAME TESTS AND ELECTRON ARRANGEMENT

NAME_____

QUESTIONS AND PROBLEMS

1. Write the electron arrangement for the following elements:

    N   _____

    Na _____

    S   _____

    Cl _____

    Ca _____

    O   _____

2. You are cooking spaghetti in water which you have salted.   You
    notice that when the water boils over that it causes cooking
    flame to turn bright yellow.   How would you explain the
    appearance of a color in the flame?

3. (a) A neutral atom has an atomic number of 80 and 45 neutrons.
         What is the nuclear symbol of the atom?

    (b) What is the nuclear symbol of a neutral atom that has
         47 electrons and 61 neutrons?

EXPERIMENT 10   COMPOUNDS AND THEIR ELEMENTS

PURPOSE

1. Identify the elements in a compound.
2. Compare some physical properties of a compound and the elements from which it was formed.
3. Use the subscripts in the formula of a compound to state the lowest whole number combination of the elements in that compound.

MATERIALS

display of elements
display of compounds
samples of iron (Fe), sulfur (S), iron and sulfur mixture (Fe + S)
    and iron(II) sulfide (FeS)
magnet
test tubes
dropper
6M HCl

KEYED OBJECTIVES IN TEXT: 3-1

DISCUSSION OF EXPERIMENT

A compound is made up of at least 2 different elements that have combined by way of chemical reactions. Almost everything you see around you is made of compounds. There are only 109 elements, but there are millions of different compunds.

A particular compound contains a definite proportion of elements. The formula of that compound indicates the number of atoms of each kind of element. For example, a molecule of the compound water has the formula H_2O. This means that two atoms of hydrogen and one atom of oxygen are always found combined in every molecule of water. Water never has any other formula.

LABORATORY ACTIVITIES

A. SOME COMPOUNDS AND THEIR ELEMENTS

A-1 Observe some physical properties of three compounds in the display of compounds. Record their formulas, names and describe their appearance.

A-2 List the elements that make up each of the compounds. Observe and record the physical properties of those individual elements.

A-3 From the formula of each compound, state the number of atoms of each element that combined to form that compound. For example, a molecule of H_2O consists of 2 atoms of hydrogen and 1 atom of oxygen.

EXPERIMENT 10   COMPOUNDS AND THEIR ELEMENTS

B.   <u>DIFFERENCES</u> <u>IN</u> <u>PROPERTIES</u> <u>OF</u> <u>A</u> <u>COMPOUND</u> <u>AND</u> <u>ITS</u> <u>ELEMENTS</u>

Place <u>small</u> <u>amounts</u> of the following in separate test tubes.
Label them A, B, C and D. (Your instructor may wish to do this
part of the experiment as a demonstration with test tubes
already prepared.)

Test tube	Contents
A	Fe
B	S
C	Fe + S (mixture)
D	FeS

B-1 Describe the physical appearance of the contents of each test
tube.

B-2 Run a magnent along the side of each test tube. If there is
any magnetic attraction, you will see particles follow the
motion of the magnet. Record your observations.

DO NOT PLACE THE MAGNET DIRECTLY INTO THE SUBSTANCE! The
attracted particles cling to the magnet make it difficult to
clean.

B-3 IN THE HOOD, add a <u>few</u> drops of 6M (6N) HCl to each test tube.
Observe any reaction in the test tube. CAREFULLY note any odor.

**CAUTION: TO SMELL A GAS, FIRST FILL YOUR LUNGS WITH FRESH
AIR, THEN USE YOUR HAND TO FAN SOME OF THE VAPORS FROM THE
TEST TUBE TOWARD YOU AND CAREFULLY SMELL.   THE GAS PRODUCED IN
THIS EXPERIMENT CAN BE TOXIC IN LARGE AMOUNTS.**

EXPERIMENT 10   COMPOUNDS AND THEIR ELEMENTS

LABORATORY RECORD

NAME_____

DATE_____

SECTION_____

A.   <u>SOME</u> <u>COMPOUNDS</u> <u>AND</u> <u>THEIR</u> <u>ELEMENTS</u>

		Compounds		
A-1	Formula			
	Name			
	Appearance			
A-2	Elements			
	Appearance			
A-3	Number of each kind of element in the formula			

B. DIFFERENCES IN PROPERTIES OF A COMPOUND AND ITS ELEMENTS

Test Tubes

		A Fe	B S	C Fe + S (mixture)	D FeS
B-1	Appearance				
B-2	Magnetic Attraction				
B-3	Reaction with HCl				
	Odor with HCl				

B-4.   Identify the contents in test tubes A, B, C, and D as

an element _____

a mixture _____

a compound _____

EXPERIMENT 10   COMPOUNDS AND THEIR ELEMENTS   NAME_____

QUESTIONS AND PROBLEMS

1.   Circle the correct answer to complete the statement.

(a) Compounds (**look like or do not look like**) the elements from

which they were formed.

(b) When elements form compounds, the characteristics of the

compounds that are formed are (**new or the same as those of the**

elements).

2.   List the number of atoms of each kind of element in the following
     formulas:

Formula   Number and Kind of Atoms in the Compound

H_2O 2 atoms H and 1 atom O _____

$CuCl_2$ _____

Al_2S_3 _____

$FeSO_4$ _____

C_4H_{10} _____

$C_6H_{12}O_6$ _____

3.   Write the formulas of the following compounds from the number
     of atoms given. The elements are listed in the order in which
     they appear in the formula.

1 atom of C and 2 atoms of O CO_2

1 atom of N and 3 atoms of H   _____

1 atom Na and 1 atom of Cl   _____

1 atom of C and four atoms of Cl   _____

2 atoms of Fe and 3 atoms of O   _____

1 atom of Ba, 1 atom of S and 4 atoms of O _____

EXPERIMENT 11 WRITING FORMULAS AND NAMES OF COMPOUNDS

PURPOSE

1. Determine the charge of an ion by observing its electron dot structure.
2. Write a correct formula and name of an ionic compound.
3. Determine the number of covalent bonds needed by elements in Groups IVA, VA, VIA, AND VIIA.
4. Write correct formulas and names for covalent compounds.

MATERIALS Paper or cards (optional)

KEYED OBJECTIVES IN TEXT: 3-1, 3-2, 3-3, 3-4, 3-5, 3-6, 3-7

DISCUSSION OF EXPERIMENT

When we consider the chemical reactivity of elements, we are primarily interested in the electrons in the highest energy level called the **valence electrons**. The octet rule tells us that atoms are most stable when they have eight electrons in the valence shell. The exceptions are the elements H and He which are stable with two electrons.

LABORATORY ACTIVITIES

PART I: IONIC CHARGES

When atoms in Groups IA, IIA, or IIIA react with atoms in Groups VA, VIA, and VIIA, they form stable octets in their valence shells by losing or gaining electrons. We can predict what this loss or gain is by observing the initial electron dot structures. An electron dot structure is written by placing dots that represent valence electrons around the symbol of the atom. Aluminum, for example, whose electron configuration is 2,8,3 has three valence electrons and an electron dot structure of

$$\cdot \overset{\bullet}{Al} \cdot$$

The aluminum atom loses its three valence electrons to become stable. The resulting aluminum ion has an electrical charge or valence of +3 and an electron configuration of 2,8 which is stable.

Al atom	Al^{3+} ion
13 p^+	13 p^+
13 e^-	10 e^-
————	————
0 charge	3 $^+$ charge

EXPERIMENT 11 WRITING FORMULAS AND NAMES OF COMPOUNDS

Elements(metals) with one, two or three valence electrons LOSE their valence electrons to form positively charged ions. Elements(nonmetals) with five, six, or seven valence electrons GAIN electrons to become stable and form negatively charged ions. In the laboratory record, complete the chart of dot structures and state their resulting ionic charges and names. The positive ions are named just like the element. The negative ions are named by changing the ending of the elemental name to -**ide**.

PART II: WRITING IONIC FORMULAS

An ionic formula represents the smallest number of positive ions and negative ions that give a charge balance of zero (0). In the laboratory record, indicate the number of each ion needed to give an overall charge balance. Write the correct formula with the number of each ion needed as a subscript. The compound is named by writing the name of the positive ion following by the name of the negative ion.

ions		formula	name
Ca^{2+}	1 needed		
Cl^-	2 needed	$CaCl_2$	calcium chloride

PART III: VARIABLE VALENCES FOR TRANSITION METALS

Many of the transition metals are capable of forming ions with more than one type of positive valence. We will illustrate variable valence with the ions of iron and copper. The names of ions with variable valences must include the particular valence.

Fe^{2+}	iron(II)	or	ferrous ion
Fe^{3+}	iron(III)	or	ferric ion
Cu^+	copper(I)	or	cuprous ion
Cu^{2+}	copper(II)	or	cupric ion

$FeCl_2$	iron(II) chloride	Cu_2O	copper(I) oxide
Fe_2S_3	iron(III) sulfide	CuS	copper(II) sulfide

Complete the formulas and names of compounds that contain transition metals.

PART IV: POLYATOMIC IONS

Polyatomic ions are groups of atoms chemically combined with an overall charge, usually negative.   Some examples are given below. A complete list of the polyatomic ions you need to know is found in your text.   Note that the endings of the polyatomic ions are different; they typically end with -ate or -ite.

NO_3^-	CO_3^{2-}	PO_4^{3-}
nitrate ion	carbonate ion	phosphate ion
HCO_3^-	SO_4^{2-}	
bicarbonate ion or hydrogen carbonate ion	sulfate ion	
NO_2^-	SO_3^{2-}	
nitrite ion	sulfite ion	

Below are some examples of ionic compounds with polyatomic ions. When more than one polyatomic ion is needed to complete the charge balance, parenthesis are used to surround the entire ion while the subscript needed to balance the charge is on the outside.   No change is made in the formula of the polyatomic ion itself.

Ions		Formula	Name
Ca^{2+}	1 needed	$Ca(NO_3)_2$	calcium nitrate
NO_3^-	2 needed		

Complete Part IV in the laboratory record.

PART V: MOLECULAR COMPOUNDS

Two different nonmetals from Groups IVA, VA, VIA, or VIIA form a compound by sharing electrons.   The resulting compound has covalent bonds and is called a **covalent compound.**   To write the formula of a covalent compound, you need to determine the number of electrons needed by each kind of atom to provide an octet.   For example, nitrogen in group V has five valence electrons.   Nitrogen atoms need 3 more electrons for an octet;   they share 3 electrons. Complete the table in the laboratory record by writing the number of electrons that each kind of atom needs to share.

79

PART VI: WRITING THE FORMULAS OF COVALENT COMPOUNDS

Formulas of covalent compounds are derived by sharing the valence electrons of the atoms until each atom acquires an octet.

(OPTIONAL) The following exercise may help you visualize the sharing of electrons in the formation of single bonded compounds.

1. Cut out or obtain 9 cards with dimensions 5 cm x 5 cm.

2. Mark the cards with the electron dot structure of the following atoms:

number	atom
1	$\cdot \overset{\bullet}{C} \cdot$
1	$\cdot \overset{\bullet}{\underset{\bullet}{N}} \cdot$
1	$: \overset{\bullet}{\underset{\bullet}{O}} \cdot$
2	$: \overset{\bullet}{\underset{\bullet}{Cl}} \cdot$
4	$H \cdot$

3. Pair up the single electrons of different atoms until an octet is formed. Remember that hydrogen atoms only need to share with one other electron to reach a stable level of 2 electrons in the first shell.

Example: Oxygen has 6 valence electrons and needs two more, and hydrogen has 1 valence electrons and needs just one. The combination of cards will look like this.

$$: \overset{\bullet\bullet}{\underset{\bullet\bullet}{O}} : H$$

                  octet

    H

The formula that represents this combination is H_2O. Write the correct formulas for the covalent compounds made up of the elements listed in the laboratory record.

EXPERIMENT 11   WRITING FORMULAS AND NAMES OF COMPOUNDS

NAME _____

DATE _____

Element	Atomic number	Electron arrangement (atom)	Dot formula	Loss or gain of electrons	Electron arrangement (ion)	Charge (ion)	Ion	Name of ion
Sodium	11	2,8,1	Na•	loss of 1e	2,8	1+	Na$^+$	sodium
Calcium								
Oxygen								
Chlorine								
Aluminum								
Fluorine								
Potassium								
Sulfur								
Magnesium								

EXPERIMENT 11 WRITING FORMULAS AND NAMES OF COMPOUNDS

B. WRITING IONIC FORMULAS   NAME_____

IONS	NUMBER NEEDED FOR CHARGE BALANCE	FORMULA	NAME
Na^+	_____		
Cl^-	_____	_____	_____
Na^+	_____		
O^{2-}	_____	_____	_____
Na^+	_____		
N^{3-}	_____	_____	_____
Mg^{2+}	_____		
Cl^-	_____	_____	_____
Mg^{2+}	_____		
O^{2-}	_____	_____	_____
Mg^{2+}	_____		
N^{3-}	_____	_____	_____
Al^{3+}	_____		
Cl^-	_____	_____	_____
Al^{3+}	_____		
O^{2-}	_____	_____	_____
Al^{3+}	_____		
N^{3-}	_____	_____	_____

NAME_____

PART III: VARIABLE VALENCES OF TRANSITION METALS

IONS	NUMBER NEEDED FOR BALANCE	FORMULA	NAME
Fe^{2+}	_____		
Br^-	_____	_____	_____
Cu^{2+}	_____		
F^-	_____	_____	_____
Fe^{3+}	_____		
S^{2-}	_____	_____	_____
Cu^+	_____		
P^{3-}	_____	_____	_____

PART IV: POLYATOMIC IONS

Mg^{2+}	_____		
NO_2^-	_____	_____	_____
Na^+	_____		
SO_4^{2-}	_____	_____	_____
Fe^{2+}	_____		
PO_4^{3-}	_____	_____	_____
$Ca2+$	_____		
HCO_3^-	_____	_____	_____

EXPERIMENT 11   WRITING FORMULAS AND NAMES OF COMPOUNDS

PART V: MOLECULAR COMPOUNDS   NAME_____

Element	Electron Dot Structure	Electrons to share
C	·Ç·	4
N		
O		
Cl		
H		

PART VI: WRITING FORMULAS OF COVALENT COMPOUNDS

Elements	Electron Dot Structure	Formula
C and H	H H :Ç: H H	CH_4
N and H		
H and Cl		
H and H		
Cl and Cl		
P and Cl		

EXPERIMENT 11: WRITING FORMULAS   NAME_____

QUESTIONS AND PROBLEMS

1. Add the correct subscript to the following compounds.

Al S   Na SO$_4$ Ca OH   K S   Mg PO$_4$ C Cl

2. Write the correct formulas for the following compounds:

sodium oxide   _____   potassium iodide _____

magnesium fluoride_____   aluminum chloride_____

copper(II) chloride_____   copper(I) oxide_____

iron(II) bromide_____   sodium carbonate_____

aluminum nitrate_____   iron(III) sulfate_____

carbon tetrachloride_____   nitrogen tribromide_____

3. Write the correct name of the following compounds:

CuO_____   N$_2$O$_4$_____

Al(NO$_3$)$_3$_____   PCl$_3$_____

FeCO$_3$_____   Na$_2$S_____

Cu(OH)$_2$_____   Ag$_2$O_____

4. Your friend wants to know what the formula FeSO$_4$ on her vitamin
    bottle means and what its name is. What would you tell her?

EXPERIMENT 12　　　**MOLES AND CHEMICAL FORMULAS**

PURPOSE

1. Identify the relationship between moles and grams of a substance.
2. Use the mole concept to convert grams to moles.
3. Determine experimentally the simplest formula for the product formed between magnesium and oxygen.

MATERIALS

Dried peas, beans, lentils   Crucible and cover
Balance   Clay triangle
Beakers   Iron ring and stand
Various compounds such as H_2O,   Bunsen burner
　NaCl, Sucrose, Na_2CO_3 , etc.   Magnesium ribbon

KEYED OBJECTIVES IN TEXT: 4-1, 4-2, 4-3

DISCUSSION OF EXPERIMENT

You probably already use certain units to represent a collection of smaller things. When you buy eggs, you purchase a dozen. How often do you actually count the eggs to see if there are 12? Most likely never, because you know that a dozen eggs is exactly 12 eggs. When you buy a ream of paper, you don't have to count each piece to know that you have 500 sheets of paper in that ream.

A chemist uses the unit called a <u>mole</u> to represent a large collection of atoms or molecules. One mole of sulfur represents 6.02×10^{23} atoms (Avogadro's number) of sulfur. The chemist also knows that this one mole of sulfur has a mass of 32.1 g.

Example of 1 mole Quantities

Substance	Number of particles	Molar Mass
1 mole Ca	6.02×10^{23} atoms Ca	40.1 g Ca
1 mole H_2S	6.02×10^{23} molecules H_2S	34.1 g H_2S
1 mole $Mg(NO_3)_2$	6.02×10^{23} formula units	148.3 g $Mg(NO_3)_2$

In the first two activities in this experiment, you will develop the idea that by weighing out a mole of a substance, you are also counting the number of atoms or molecules of that substance. Then you will actually measure out and look at a mole of some substances.

In the last activity, you determine the simplest chemical formula of a compound formed from its elements. Then, you will use your ideas of moles and grams to do calculations.

EXPERIMENT 12   MOLES AND CHEMICAL FORMULAS

LABORATORY ACTIVITIES

PART I: ATOMS AND MOLES

A.   Counting Atoms

A-1 Using the laboratory balance, determine the mass in grams of ten dried peas or beans.

A-2 Calculate the average mass of a pea or bean.

$$\frac{\text{total mass}}{\text{10 items}} = \text{average mass(g)/item}$$

A-3 Calculate the mass that would contain 50 of the peas or beans. DO NOT COUNT.

$$50 \text{ items } x \ \frac{\text{average mass (g)}}{\text{item}} = \text{mass(g)}$$

A-4 Weigh out the quantity you calculated. Now count the actual number of peas or beans. Record that number.

A-5 Calculate the mass in grams that will contain 5000 peas or beans. (DON'T TRY TO COUNT THEM ALL!) As the numbers become larger, you will see that it becomes difficult to count the individual peas or beans.

B. "Seeing" a Mole"

In the laboratory, there is a display of several 1 mol quantities of substances. Write down the formula of the compound. Calculate the mass of compound that is present in each of the samples by determining the molar mass of each.

PART II: FINDING THE SIMPLEST CHEMICAL FORMULA

The simplest formula of a compound represents the lowest whole number ratio of the elements in a single unit of that compound. In this experiment, you will weigh out in grams a certain amount of the starting material. You will need to use the atomic mass of each element (from the periodic table). For quantitative experiments like this, be sure to use clean equipment and to weigh carefully. The experiment is not difficult, but it does require careful attention to the skills you have learned.

C.   Determination of Mass

C-1 Weigh a clean, dry crucible and cover. (A procelein crucible may have stains in the porcelein that cannot be removed.) Record the combined mass.

C-2 Obtain a length of magnesium ribbon that has a mass between
0.50 and 0.80 g. You may need to remove the oxide coating by
polishing the ribbon with steel wool.   Wind the ribbon into a
loose coil and place in the bottom of the crucible. Replace the
cover. Weigh the ribbon, crucible and cover as accurately as
you can. Record.

C-3 Place the crucible, ribbon and cover on the clay triangle. Set
the cover ajar and begin to heat. The magnesium will begin to
react with the oxygen in the air. When the magnesium ribbon
begins to smoke or bursts into flame, place the cover over the
crucible using your tongs.   Avoid looking directly at the
bright flame of the burning magnesium. Open up the crucible by
setting the cover ajar again. Continue this process until the
magnesium no longer produces smoke or glows brightly.   Set the
cover ajar and heat strongly for 10 minutes.   Then let the
crucible and its contents cool.   See Figure 12-1.

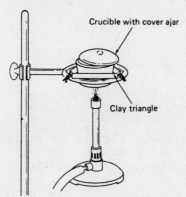

Figure 12-1   Crucible and cover on clay triangle set on
an iron ring.

At this point, the ribbon in the crucible should be a whitish-
gray ash. Since there is nitrogen in the air, some of the
magnesium will be combined with the nitrogen.   To get rid of
the nitrogen, let the contents cool. Then, use a dropper to
carefully add 15-20 drops of water to the cooled contents.
Return the crucible, contents and cover to the clay
triangle. Set the cover slightly ajar and heat gently for
about 5 minutes to drive off the water and ammonia. Heat
strongly for 5 more minutes.

$$Mg_3N_2 \ + \ 3H_2O \quad \longrightarrow \quad 3MgO \ + \ 2NH_3(g)$$

Place the crucible, cover and contents on a heat resistant
pad and let cool. Weigh the cooled crucible, cover and
contents as accurately as you can. Record.

EXPERIMENT 12   MOLES AND CHEMICAL FORMULAS

D.   Calculating the Simplest Formula

D-1   Calculate the mass of the magnesium ribbon.   Record.

D-2   Calculate the mass of the oxide product.

D-3   Calculate the mass of oxygen that reacted with the magnesium
      ribbon.   (Subtract the mass of the ribbon (D-1) from the
      mass of the oxide product (D-2).   Record.

D-4   Calculate the moles of magnesium.   Record.

$$\text{mole Mg} = \text{mass Mg} \times \frac{\text{1 mol Mg}}{\text{24.3 g}}$$

D-5   Calculate the moles of oxygen that reacted with the magnesium
      ribbon.   Record.

$$\text{mole O} = \text{mass O} \times \frac{\text{1 mol O}}{\text{16.0 g}}$$

D-6   Which is the smallest number of moles, Mg or O?

D-7   Divide the number of moles of magnesium and the number of
      moles of oxygen by the smaller number of moles.   Record.

D-8   Usually the mole ratio should be small numbers such as 1:1,
      1:2, 2:1, 1:3, 2:3, and so on.   Round off your experimental
      values to whole numbers.   Write the simplest formula for the
      oxide product.

D-9   Using the masses of magnesium and oxygen obtained experimentally,
      calculate the % Mg in the product.

LABORATORY RECORD   NAME_____
                                           DATE_____
PART I: ATOMS AND MOLES   SECTION_____

A. Counting Atoms

A-1   Mass of 10 dried _____   = _____
                        (peas, beans, etc.)
A-2   Average mass (g/item)   _____
      Calculation:

A-3   Mass that contains 50 peas or beans   _____
      Calculation:

A-4   Actual number present

      Did you have 50 peas(beans) or close?

      What do you think is meant by the phrase, "counting by
      weighing"?

A-5   Mass that contains 5000 peas or beans   _____
      Calculations:

B."SEEING " A MOLE

Formula	Molar mass

PART II: FINDING THE SIMPLEST CHEMICAL FORMULA

C. <u>DETERMINATION</u> <u>OF</u> <u>MASS</u>

C-1 Mass of empty crucible and cover   _____

C-2 Mass of crucible, magnesium ribbon and cover   _____

C-3 Mass of crucible, oxide product, and cover   _____

D. <u>CALCULATING</u> <u>THE</u> <u>SIMPLEST</u> <u>FORMULA</u>

D-1 Mass of magnesium ribbon   _____

D-2 Mass of oxide product   _____

D-3 Mass of oxygen in the product   _____

D-4 Moles of magnesium   _____
     Calculations:

D-5 Moles of oxygen   _____
     Calculations:

D-6 Smallest number of moles

D-7 Calculation of mole ratios:

<u>moles</u> <u>of</u> <u>magnesium</u>   =   _____moles Mg
smaller number of moles

<u>moles</u> <u>of</u> <u>oxygen</u>   =   _____moles O
smaller number of moles

D-8 Simplest formula   _____

D-9 Percent Mg in product   _____

EXPERIMENT 12   MOLES AND CHEMICAL FORMULAS

QUESTIONS AND PROBLEMS   NAME_____

1. Using your rules for writing the formulas of ionic compounds,
    write the correct ionic formula for magnesium oxide.

2. What is the %Mg for the magnesium oxide?

3. How does the calculated value in question 2 compare to your
    experimental value in D-9?

4. Explain the effect on your answer for %Mg (low, high or no
    change) if the following errors in experimental procedure occur:

    a. Some magnesium ribbon did not completely react.

    b. A portion of the oxide product was spilled before the final
        weighing.

    c. Some of the water added in step C-3 did not completely
        boil off.

93

EXPERIMENT 13   PHYSICAL AND CHEMICAL CHANGES

PURPOSE

1. Observe physical and chemical changes associated with physical processes and chemical reactiOns.
2. Give evidence for the occurrence of a chemical reaction.
3. Write a balanced equation for a chemical reaction.

MATERIALS
test tubes
test tube rack
chemicals: solids $Na_2CO_3(s)$, Mg ribbon, $Zn(s)$, $CuSO_4.5H_2O(s)$, $NH_4NO_3(s)$, $CaCl_2$ (anhydrous), ice

solutions $BaCl_2$, Na_2SO_4, $CuSO_4$, $NaCl$, $AgNO_3$, Na_3PO_4, 1 M HCl

thermometer
glass stirring rod

KEYED OBJECTIVES IN TEXT: 4-4, 4-5, 4-6

DISCUSSION OF EXPERIMENT

This is an experiment in making observations. You will be looking at many chemical reactions as you proceed. You are to make as many observations of substances as you can before and after they undergo a reaction.   Describe their physical properties carefully. Making observations will help you learn about physical and chemical changes. A physical change has occurred if the nature of the substance is retained. For example, steam is water in the form of a gas.   A temperature change will convert the gaseous form of water, steam, back to liquid. When a change occurs that is associated with an alteration in the nature of the substance, then a chemical change has occurred.

Physical Changes	Chemical Changes
change in state	When the following are not associated
change in size	with a change of state, they represent a
tearing	chemical change:
breaking	formation of a gas (bubbles)
grinding	formation of a solid (precipitate)
separating	disappearance of a solid (dissolving)
mixing	change in color
	evolution or absorption of heat
	a change in the pH (acidity)
	a change in the temperature

EXPERIMENT 13   PHYSICAL AND CHEMICAL CHANGES

LABORATORY ACTIVITIES

For all of the following experiments, use small quantities. For
solids, use the amount of compound that will fit on the tip of a
spatula or small scoop. Carefully pour small amounts of liquids
into your own beakers and other containers.   Measure out 3-4 mL of
water in a test tube similar to those you will be using. Use this
volume as a reference level. Judging by eye, use about the same
level of solution for each of the experiment.

**DO NOT PLACE DROPPERS OR STIRRING RODS INTO ANY REAGENT BOTTLES.
NEVER RETURN A CHEMICAL TO ITS ORIGINAL CONTAINER.   DISCARD IN
THE APPROPRIATE MANNER INDICATED BY YOUR INSTRUCTOR.**

Read directions carefully. Match labels on bottles and containers
to the names and concentrations of the compounds given in your
directions before you proceed with that chemicals. Kept your desk
neat and orderly. Label each of your containers with the formulas
of the chemicals as you remove them from the original laboratory
bottle.

WEAR GOGGLES DURING EXPERIMENTAL PROCEDURES.

Record your observations of the physical properties of the
substance before and after any reaction for each of the following
experiments on the laboratory record. USE COMPLETE SENTENCES TO
DESCRIBE YOUR OBSERVATIONS.   State whether an observation
represents a physical or chemical change.   The instructions will
include a net (unbalanced) equation for the reaction. Write the
complete, balanced equations for the reactions as indicated on the
laboratory record.

A.   Obtain a small strip of magnesium ribbon. Record your
     observations of the ribbon. Holding the end of the strip using
     a pair of <u>tongs</u>, place the magnesium ribbon in the flame of
     the Bunsen burner. SHIELD YOUR EYES when the ribbon ignites
     and remove the ribbon from the flame.   Record your final
     observations.   Write the balanced equation for the reaction.

     net $Mg(s)$ + $O_2(g)$ $\longrightarrow$ $MgO\ (s)$
     equation

B.   Place 3-4 mL of 0.1M $CuSO_4$ solution into each of two test
     tubes.   Add a small piece of Zn metal to one of the samples.
     Observe the solutions with and without the metal at 1 minute,
     15 minutes and 20 minutes.   Record your observations.
     Write the balanced equation for the reaction.

     net $Zn(s)$ + $CuSO_4$ $\longrightarrow$ $Cu\ (s)$ + $ZnSO_4$
     equation

C. Place a small amount of solid $CuSO_4 \cdot 5H_2O(s)$ in a test tube.
   Describe the $CuSO_4 \cdot 5H_2O$ crystals. Holding the test tube with a
   test tube holder, heat gently as you move the test tube through
   the flame. See Figure 13-1 for the proper way to heat a
   substance in a test tube.

DO NOT HEAT A TEST TUBE IN ONE SPOT.   KEEP THE TEST TUBE MOVING.

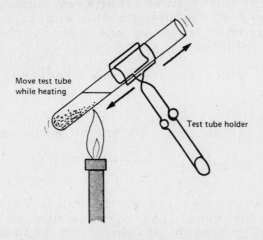

Move test tube
while heating

Test tube holder

Figure 13-1   Heating a sample in a test tube.

Describe the appearance of both the heated crystals and the
upper part of the test tube.   Return the test tube to your test
tube rack and let it cool. Write a balanced equation for the
reaction.

net
equation $\qquad CuSO_4 \cdot 5H_2O \longrightarrow \qquad CuSO_4 \quad + \quad H_2O$

When the test tube and contents have cooled, add a few drops of
water to the test tube.   Record your observations.

net
equation $\qquad CuSO_4(s) \quad + \quad H_2O(l) \longrightarrow \quad CuSO_4 \cdot 5H_2O$

EXPERIMENT 13 PHYSICAL AND CHEMICAL CHANGES

D-1 Place 5 mL of water in a test tube. Use a thermometer to determine its temperature. Record. Add a scoop of solid NH_4NO_3 to the water. Stir and record the temperature of the solution.

net
equation $\quad NH_4NO_3(s) \xrightarrow{\;H_2O\;} NH_4^+(aq) \quad + \quad NO_3^-(aq)$

$\qquad\qquad\qquad\qquad\qquad\qquad$ (aq) = dissolved in water

D-2 Place 5 mL of water in another test tube. Use a thermometer to record the temperature of the water. Add a scoop of solid $CaCl_2$ (anhydrous). Stir and record the temperature of the resulting solution.

net
equation $\quad CaCl_2(s) \longrightarrow Ca^{2+}(aq) \quad + \quad 2Cl^-(aq)$

E. Place some ice chips or an ice cube on a watch glass . Record your observations at 1, 10 and 30 minutes.

net
equation $\quad H_2O(s) \longrightarrow H_2O(l)$

F. In the following experiments, you will combine two solutions and look for resulting changes due to a chemical reaction. Use 2-3 mL of each of the 0.1M solutions. Describe the appearance of the solutions before and after they are mixed. Balance any equations that are given.

F-1 $BaCl_2$ and Na_3PO_4

net
equation $\quad BaCl_2 \quad + \quad Na_3PO_4 \longrightarrow Ba_3(PO_4)_2(s) \quad + \quad NaCl$

F-2 $BaCl_2$ and Na_2SO_4

net
equation $\quad BaCl_2 \quad + \quad Na_2SO_4 \longrightarrow BaSO_4(s) \quad + \quad NaCl$

F-3 NaCl and $AgNO_3$

$\qquad\qquad$ CAUTION: CARE MUST BE TAKEN WITH POURING OUT $AgNO_3$. DO NOT SPILL IT ON YOUR SKIN - IT LEAVES A PERMANENT STAIN.

net
equation $\quad NaCl \quad + \quad AgNO_3 \longrightarrow AgCl(s) \quad + \quad NaCl$

G. Place a small amount of solid Na_2CO_3 in a test tube. With a medicine dropper, add dropwise, 2-3 mL of 1M HCl. Record your observations. Balance the equation for the reaction.

net
equation $\quad Na_2CO_3(s) + HCl \longrightarrow CO_2(g) \quad + H_2O + NaCl$

Experiment	Observations	Type of Reaction	Balanced Equation
A	Initial: Final:		
B	1 minute 15 minutes 30 minutes		
C	Before heating: After heating: After H_2O added:		

EXPERIMENT   13   PHYSICAL AND CHEMICAL CHANGES   NAME _____

Experiment	Observations	Type of	Balanced Equation
D-1			
D-2			
E.	1 minute		
	10 minutes		
	30 minutes		

EXPERIMENT 13   PHYSCIAL AND CHEMICAL CHANGES   NAME _____

Experiment	Observations	Type of Reaction	Balanced Equation
F-1			
F-2			
F-3			
G			

EXPERIMENT 13   PHYSICAL AND CHEMICAL CHANGES

QUESTIONS AND PROBLEMS

1. What are four observations that might indicate that a chemical reaction has occurred.

2. What evidence of a chemical reaction might you see in the following?

   a. fizzing of an Alka-Seltzer tablet dropped in a glass of water

   b. bleaching a stain

   c. burning a match

   d. rusting of an iron nail

3. Balance the following equations:

   a. $Mg(s)$ + HCl $\longrightarrow$ $H_2(g)$ + $MgCl_2$

   b. $Al(s)$ + $O_2(g)$ $\longrightarrow$ $Al_2O_3(s)$

   c. $NaOH$ + H_3PO_4 $\longrightarrow$ Na_3PO_4 + H_2O

   d. Fe_2O_3 + H_2O $\longrightarrow$ $Fe(OH)_3$

   e. $Ca(OH)_2$ + HNO_3 $\longrightarrow$ $Ca(NO_3)_2$ + H_2O

4. Write an equation for the following reactions and balance:

   a. potassium and oxygen gas react to form potassium oxide

   b. sodium and water react to form sodium hydroxide and hydrogen gas

   c. iron and oxygen gas react to form iron(III) oxide.

102

EXPERIMENT 14   FORMULA OF A HYDRATED SALT

PURPOSE

1. Determine the percent water in a hydrated salt.
2. Calculate the number of moles of water in a hydrated salt.
3. Write the correct formula of a hydrated salt.

MATERIALS

hydrated $MgSO_4$
crucible
clay triangle

DISCUSSION OF EXPERIMENT:

Certain solid forms of salts contain some water molecules loosely bonded to a formula unit of the salt. These solids are called hydrates, and the molecules of water are the water of hydration.

Since the number of water molecules is specific for each kind of hydrate, water is included in the formula of the compound. For example, the hydrate formula of copper sulfate includes the formula copper sulfate followed by a dot and the number of water molecules. One formula unit of copper sulfate is combined with five water molecules, and the formula is:

$$CuSO_4 \cdot 5H_2O$$

Since the water molecules are held by weaker attractive forces than the ionic bonds of the compound, the water molecules can be removed by heating. A hydrate that loses its water molecules is called an anhydrate.

$$CuSO_4 \cdot 5H_2O \xrightarrow{\text{heat}} CuSO_4 \quad + \quad 5H_2O(g)$$

| hydrate | anhydrate | water of hydration |

Examples of some other hydrates include $CaCl_2 \cdot 2H_2O$, $CoCl_2 \cdot 6H_2O$, and $Na_2CO_3 \cdot 10H_2O$.

LABORATORY ACTIVITIES

A. <u>DEHYDRATION</u>

A-1   Obtain a clean, dry crucible. Weigh to the nearest 0.01 g and
      record its mass.

A-2   Place about 4 to 5 g of the magnesium sulfate hydrate in the
      crucible, and weigh the crucible and its contents. Record the
      mass.

A-3   Set the crucible and hydrate on a clay triangle that is resting
      in a iron ring.   Heat gently for 2 minutes, and then heat
      strongly for about 8-10 minutes more. Do not heat so strongly
      that the crucible turns red.   Remove heat and let the crucible
      cool.   Weigh the crucible and its contents.

      You may wish to reheat the crucible and contents to make sure
      that you have removed all the water of hydration. Reheat for
      4-5 minutes and cool again. Reweigh the crucible and contents.
      If the mass is within 0.01 g of the mass in A-3, then you have
      completely removed the water. If not, you may reheat again
      until you have agreement between final weighings. Use the
      final weighing for your calculations.

B.   <u>CALCULATIONS</u>

B-1   Calculate the mass of the hydrated magnesium sulfate sample
      before heating.

B-2   Calculate the mass of the dehydrated magnesium sulfate
      sample after heating.

B-3   The difference in the mass of the hydrate and the anhydrate
      is the mass of water lost during the heating. Calculate
      the grams of water lost.

B-4   Calculate the percent water of hydration in the hydrate.

$$\frac{\text{g water lost}}{\text{g hydrate sample}} \quad \times \ 100 \quad = \quad \% \ H_2O$$

B-5   Calculate the moles of H_2O lost by the hydrate.

$$g \ H_2O \ lost \ \times \ \frac{1 \ mole \ H_2O}{18.0 \ g \ H_2O} \ = \ moles \ H_2O$$

B-6   Calculate the moles of an hydrate in the sample(the formula weight of $MgSO_4$ anhydrate is 120.4).

$$g \ MgSO_4(anhydrate) \ \times \ \frac{1 \ mole \ MgSO_4}{120.4 \ g \ MgSO_4} \ = \ moles \ MgSO_4(anhydrate)$$

B-7   Determine the ratio of moles of water of hydration per mole of magnesium sulfate anhydrate.

$$\frac{mole \ H_2O}{mole \ MgSO_4(anhydrate)} \ = \ mole \ H_2O/1 \ mole \ MgSO_4(anhydrate)$$

B-8   Round off the value in B-7 to its nearest whole number(n). Write the formula of the hydrated magnesium sulfate.

$$MgSO_4 \cdot nH_2O$$

EXPERIMENT 14   FORMULA OF A HYDRATE   NAME_____

DATE_____

LABORATORY RECORD   SECTION_____

A. <u>DEHYDRATION</u>

A-1   Mass of crucible   _____

A-2   Mass of crucible and salt (hydrate)   _____

A-3   Mass of crucible and salt (anhydrate)

_____
1st weighing

_____
2nd weighing
(if needed)

_____
3rd weighing
(if needed)

B. <u>CALCULATIONS</u>

B-1   Mass of hydrate   _____

B-2   Mass of anhydrate   _____

B-3   Mass of water of hydration   _____

B-4   % water of hydration   _____
Calculations:

B-5   Moles of water of hydration   _____
Calculations:

B-6   Moles of salt(anhydrate)   _____
Calculations:

B-7   Moles of H_2O/ 1 mole anhydrate $MgSO_4$ _____
Calculations:

B-8   Formula of hydrate   _____

106

QUESTIONS AND PROBLEMS

1. Using the formula you obtain in B-8, write the equation for the dehydration of the hydrate when it is heated.

2. Write an equation for the dehydration of the following hydrates:

    $BaCl_2 \cdot 2H_2O$

    $CoCl_2 \cdot 6H_2O$

3. What is the % water of hydration for the hydrate $Na_2CO_3 \cdot 10H_2O$?

4. Will the experimental %H_2O be too high ot too low if the hydrate sample is not heated sufficiently to drive off all of the H_2O?

5. Using the actual formula of the hydrated $MgSO_4$ (see instructor), calculate the true value for the % H_2O of hydration.

6. Using the actual formula of hydrated $MgSO_4$ (see instructor), calculate the percentage error.

    % error= $\dfrac{\text{difference between experimental \%}H_2O \text{ and actual \%}H_2O}{\text{actual \%}H_2O}$

EXPERIMENT 15   DETECTING RADIATION

PURPOSE

1. Observe the use of a Geiger-Mueller radiation detection tube.
2. Determine the effect of shielding materials, distance, and time on radiation.

MATERIALS

Gieger-Mueller radiation detection tube
Radioactive sources
Meterstick
Shielding materials such as lead, paper, glass, cardboard, etc.

KEYED OBJECTIVES IN TEXT: 5-1, 5-2, 5-3, 5-4, 5-5, 5-6, 5-7, 5-8

DISCUSSION OF EXPERIMENT

Radioactivity occurs through a change in the components of the nucleus of an unstable atom. When such a change takes place, alpha particles, beta particles, or gamma rays are emitted from the nucleus. These particles and rays are collectively called nuclear radiation.

       Radiation can pass through the cells of the body and damage them. You can protect yourself by using shielding materials, by limitng time spent in radioactive areas, and by keeping a reasonable distance from the radioactive source.

       To detect radiation a Geiger-Mueller tube is used. Radiation passes through the gas held within the tube, producing ion pairs. These charged particles emit bursts of current that are converted to flashes of light and audible clicks. In this experiment your teacher will demonstrate the use of the Geiger-Mueller tube to test the effects of shielding, time, and distance. RADIOACTIVITY IS HAZARDOUS: FOLLOW YOUR INSTRUCTOR'S DIRECTIONS CAREFULLY.

LABORATORY ACTIVITIES: (TEACHER DEMONSTRATION)

PART I: RADIATION DETECTION AND BACKGROUND COUNT

A.   BACKGROUND COUNT

A-1 Set the radiation counter at proper voltage for operating level. Let it warm up for 5 minutes. Remove all sources of radiation near the counter. Count the radiation present in the room by operating the counter for three 1-minute intervals, recording the counts for each 1-minute interval.

EXPERIMENT 15   DETECTING RADIATION

A-2 Total the counts and divide by 3. This value represents the background radiation in counts/minute. This background count is the natural level of radiation that constantly surrounds and strikes us. For the rest of this experiment, the background radiation _must_ _be_ _subtracted_ to give the radiation from the radioactive source alone.

B.   RADIOACTIVE SOURCE

B-1 Obtain a radioactive source. Record the radioactive source in the laboratory record.

B-2 Place the radioactive sample near the detection tube. Operate for 1 minute. Record the counts/minute after subtracting the background count.

PART II: EFFECT OF SHIELDING, TIME AND DISTANCE

C.   SHIELDING

C-1 Place various shielding materials between the radioactive source and the detection tube. Do not vary the distance of the source from the tube. (See Figure 15-1.) Record the type of shielding used.

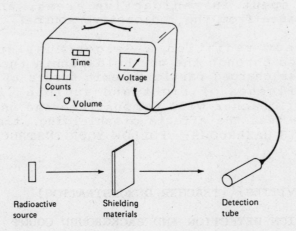

Figure 15-1 Testing shielding effects.

C-2 Record the counts/minute with each type of shielding.

C-3 Answer the questions in the laboratory record.

EXPERIMENT 15   DETECTING RADIATION

D. <u>TIME</u>

D-1   The greater the length of time you spend near a radioactive
       source, the greater the amount of radiation you receive.
       Keeping the distance constant, record the total counts for a
       radioactive source for 1, 2 and 5 minutes. Subtract background
       radiation from each.

D-2   Use the results from D-1 to calculate the counts for 10, 30
       and 60 minutes.

E. <u>DISTANCE</u>

E-1   By doubling your distance from a radioactive source, you
       receive 1/2 x 1/2 or 1/4 the intensity of the radiation. See
       Figure 15-2. Place a radioactive source 1 m (100 cm) from the
       tube.   Calculate the counts/minute at 100 cm.

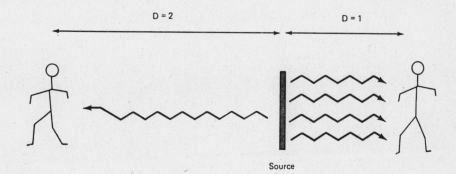

Figure   15-2   The effect of radiation lessens as the
                  distance from the source increases.

E-2   Decrease the distance of the radioactive source from the
       detection tube to 75 cm, 50 cm 25 cm, 10 cm. Stop if the
       counts get too great for the operating level of the counter.
       Record the counts/minute at each distance.

E-3   Graph this data. Graph paper can be found in the back of this
       lab book. Place the counts/minute on the vertical axis and
       the distance from the source on the horizontal axis.   See
       Appendix C for help in drawing graphs.

E-4   Using the graph, obtain the counts/minute at distances of 20
       cm and 40 cm.

E-5   Calculate the ratio of the counts at these distances(20 and 40
       cm) to find the increase in radiation with a halving of the
       distance to the source.

EXPERIMENT 15   DETECTING RADIATION

LABORATORY RECORD

NAME_____

DATE_____

SECTION_____

PART I: RADIATION DETECTION AND BACKGROUND COUNT

A. BACKGROUND COUNT

A-1 Counts during minute 1 _____

minute 2 _____

minute 3 _____

A-2 Total counts _____

Average background count _____ counts/minute

B. RADIOACTIVE SOURCE

B-1 Type of radioactive source _____
       (alpha particles, beta particles, gamma rays)

B-2 _____ - _____ = _____
        counts/min   background   cpm for source

PART II: EFFECT OF SHIELDING, TIME AND DISTANCE

C. SHIELDING

Background Count _____

(1) Type of Shielding   (2) Counts/minute

_____   _____

_____   _____

_____   _____

_____   _____

_____   _____

C-3 Which type of shielding would offer the best protection from
     radiation?

D.   TIME

Background Count _____

D-1 Number of Minutes   Counts

          1   _____

          2   _____

          5   _____

D-2   10 (calculated) _____

          30 (calculated) _____

          60 (calculated) _____

E.   DISTANCE

E-1 Distance from Radioactive Source   E-2 Counts/minute

          _____cm   _____

          _____cm   _____

          _____cm   _____

          _____cm   _____

          _____cm   _____

          _____cm   _____

E-3 GRAPH

     Prepare a graph with **counts/min** on the vertical axis and
     distance from source on the horizontal axis. Use the graph to
     answer the following parts of the experiment.

E-4 Counts/min at 40 cm = _____   Counts/min at 20 cm = _____

E-5 Ratio: Counts/min at 20 cm   = _____
              Counts/min at 40cm

114

QUESTIONS AND PROBLEMS

1. Write the symbols for of the following types of radiation:

    a. alpha particle _____   b. beta particle _____

    c. gamma ray   _____

2. List some shielding materials adequate for protection from:

    a. alpha particles _____

    b. beta particles _____

    c. gamma rays _____

3. Complete the nuclear equations by filling in the correct symbols:

$$^{27}_{13}\text{Al} \quad + \quad \underline{\hspace{2cm}} \quad \longrightarrow \quad ^{24}_{11}\text{Na} \quad + \quad ^{4}_{2}\text{He}$$

$$^{125}_{53}\text{I} \quad + \quad ^{0}_{-1}\beta \quad \longrightarrow \quad \underline{\hspace{2cm}}$$

$$^{96}_{40}\text{Zr} \quad + \quad \underline{\hspace{3cm}} \quad \longrightarrow \quad ^{1}_{0}\text{n} \quad ^{99}_{42}\text{Mo}$$

4. Describe a radioisotope used in nuclear medicine:

5. Write a paragraph for a patient handbook comparing the
    advantages and disadvantages of taking a radioisotope such as
    radioactive iodine for a diagnostic study.

EXPERIMENT 16   **CHANGES OF STATE**

PURPOSE

1. Collect data while heating and cooling a substance.
2. Prepare graphs of heating and cooling curves.
3. Use the graphs to detemine melting and boiling points.

MATERIALS

stirring apparatus(two-holed stopper, wire stirrer, and
    thermometer)
test tube to hold stirring apparatus
400-mL beaker
wire screen
a timer or a watch with a second hand
ice
water
phenylsalicylate ("salol")

KEYED OBJECTIVES IN TEXT: 6-1, 6-2, 6-3, 6-4

DISCUSSION OF EXPERIMENT

The molecules, atoms, or ions of a liquid are in constant motion.
When the liquid is heated, some molecules gain enough energy to
escape from the liquid. When the molecules begin to form bubbles
within the liquid, we say that the liquid is <u>boiling</u>. If the
liquid is cooled, the particles move more slowly, eventually
arranging themselves into a regular pattern characteristic of the
solid structure of that substance. This change of state is called
<u>freezing</u>.

     A change of state becomes more obvious when the temperature
of the substance is plotted against time. When a liquid boils, the
temperature remains constant and a horizontal line or plateau
appears on the graph. The temperature at which that plateau occurs
is the boiling point of the liquid. It will not increase any
further as long as the liquid continues to boil.   A plateau also
appears on a cooling curve of a liquid indicating the change of
state from liquid to solid which occurs at the freezing point.

LABORATORY ACTIVITIES

PART I: HEATING CURVE FOR WATER

A. <u>DATA</u> <u>FOR</u> <u>A</u> <u>HEATING</u> <u>CURVE</u>

Fill a 400-mL beaker about half-full with ice. Add about 100 mL of
water to the ice and allow the mixture to reach a constant
temperature. Set the beaker and ice on a wire screen that has been

placed on the ringstand. Be sure you have adjusted the ringstand
so that the flame from the Bunsen burner will touch the bottom of
the wire screen and beaker of ice.   Place a thermometer in a clamp
so the bulb rests about midway in the ice-water slurry. Have a
stirring rod ready to stir the ice and water as you proceed.

With a timer set at 0 minutes, stir the ice-water slush and
record its temperature. Begin heating as you start timing. Record
the temperature every 30 seconds (or 0.5 minutes). Be sure to
continue stirring to equilibrate the temperature between the
bottom, sides and middle of the sample of water. Mark the
temperature when the last piece of ice melts.

Continue to heat even when the water is boiling. Mark the
temperature at which the water boils. Make sure you continue to
record the temperature for an additional 3 or 4 minutes even though
it remains constant after vigorous boiling has occurred.

B.   DRAWING A HEATING CURVE

B-1   On the graph paper provided at the back of this lab book,
      prepare a graph with temperature units (evenly spaced) on
      the vertical axis, and the time units on the horizontal
      axis(evenly spaced). Use most of the space, making the range
      of temperature and time fit the data.   Refer to
      Appendix C for more directions on preparing a graph.
      Draw a smooth, solid line through the points.   When the
      graph goes through melting and boiling points, the slope goes
      to zero, and a flat line will result.   See Figure 16-1.

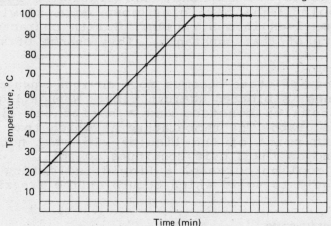

Figure 16-1   Example of a graph for plotting heating
                and cooling curves.

B-2   Title the graph and indicate the states of matter(solid or
      liquid) and the change(s) (melting or boiling) of state.
      Indicate the melting point and the boiling point for water.

EXPERIMENT 16   CHANGES OF STATE

PART II: A COOLING CURVE FOR PHENYLSALICYLATE, "SALOL"

C.   DATA FOR A COOLING CURVE

The substance in this experiment, phenylsalicylate (salol), is a solid at room temperature. Since this substance is difficult to clean out of the test tube, your instructor may have some test tubes and stirring set ups already prepared. If so, obtain one of these set ups from your instructor. DO NOT TRY TO REMOVE THE THERMOMETER. IT IS FROZEN IN THE SOLID SUBSTANCE AND WILL BREAK.

If a set up is not available, prepare one by placing phenylsalicylate in the bottom of a dry test tube to a depth of about 8 cm. The test tube must fit the stirring apparatus. This consists of a double-holed slotted stopper with a thermometer placed in the slotted hole. The stirring wire goes through the other hole. BE EXTREMELY CAREFUL IF YOU MUST PLACE THE THERMOMETER IN THE STOPPER. HOLD THE THERMOMETER IN THE MIDDLE WITH A PAPER TOWEL. WET THE STOPPER AND VERY SLOWLY BEGIN WORKING THE THERMOMETER THROUGH. The bulb of the thermometer must eventually be centered in the liquid after the phenylsalicylate melts. The loop of the stirring wire goes around the thermometer. See Figure 16-2.

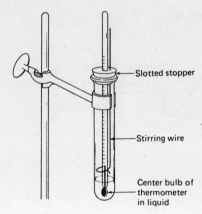

Figure 16-2 Stirring apparatus for freezing point determination.

Place the test tube set up and contents in a hot water bath and heat it to a temperature of 60-65° a temperature that is above the melting point. Remove the flame, leaving the apparatus in the hot water until the salol has completely melting.

When you are ready to start timing, remove the test tube and its contents. Record this temperature for minutes = 0. While moving the wire stirrer, continue to record the temperature every minute.

Return the stirring apparatus and contents to the instructor when you are finished with this experiment.

D. GRAPHING THE COOLING CURVE

D-1 Plot the results of the cooling curve on the graph paper
     provided in the laboratory record.   Label each axis,
     temperature on the vertical axis and time minutes on the
     horizontal axis.   Indicate the areas of liquid and solid, and
     the freezing point for the substance.

D-2 Sometimes, before freezing can occur, the temperature of a
     liquid goes below its actual freezing piont. This is called
     "supercooling".   When the particles are finally arranged to
     form the solid structure, freezing occurs, and the temperature
     rises to a constant value as freezing continues.   This is the
     freezing point.   When all of the substance has frozen, the
     temperature will begin to drop again as the solid continues to
     cool.   However, it cannot cool any further than room
     temperature. Mark the area of supercooling if present.   See
     Figure 16-3.

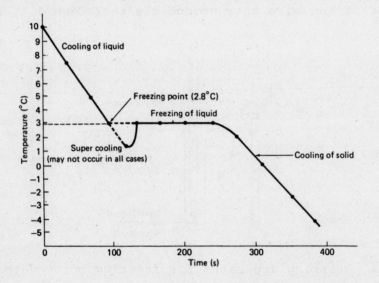

Figure 16-3 A typical cooling curve.

LABORATORY RECORD   NAME_____

                                        DATE_____

PART I: HEATING CURVE FOR WATER   SECTION_____

A.   DATA FOR THE HEATING CURVE

Time(min)	Temperature(^{o}C)	Time(min)	Temperature(^{o}C)
0.0 min	_____	_____	_____
0.5 min	_____	_____	_____
_____	_____	_____	_____
_____	_____	_____	_____
_____	_____	_____	_____
_____	_____	_____	_____
_____	_____	_____	_____
_____	_____	_____	_____
_____	_____	_____	_____
_____	_____	_____	_____
_____	_____	_____	_____

B. Graph the cooling on the graph paper provided in the
    laboratory manual. Be sure to mark the states of
    matter (liquid and solid), the change(s) of state (melting
    and freezing) and the melting point and the boiling point.

    What is the melting point of water?   _____

    What is the boiling point of water?_____

EXPERIMENT 16   CHANGES OF STATE

LABORATORY RECORD   NAME_____

PART II: COOLING CURVE FOR PHENYSALICYLATE, "SALOL"

C.   DATA

Time(min)	Temperature(^{O}C)	Time(min)	Temperature(^{O}C)
0	_____	_____	_____
1	_____	_____	_____
2	_____	_____	_____
_____	_____	_____	_____
_____	_____	_____	_____
_____	_____	_____	_____
_____	_____	_____	_____
_____	_____	_____	_____
_____	_____	_____	_____
_____	_____	_____	_____
_____	_____	_____	_____
_____	_____	_____	_____
_____	_____	_____	_____
_____	_____	_____	_____
_____	_____	_____	_____

D.   Graph a cooling of phenylsalicylate placing the temperature on the
     vertical axis and the time (minutes) on the horizontal axis.
     Identify the states (solid and liquid) and the change of state.
     Indicate supercooling and the freezing point.

122

QUESTIONS AND PROBLEMS

1. How would a burner with a larger flame affect the time required
   to reach the boiling point of water?

   How would the boiling point be affected?

2. How did you determine the freezing point of the "salol"?

3. How many minutes did the "salol" require to completely freeze.

PURPOSE

1. Experimentally determine the specific heat of a metal or other substance.
2. Observe the change in temperature during the solution process.
3. Use a temperature change to calculate energy change in a reaction.
4. Identify a reaction as endothermic or exothermic.

MATERIALS
styrofoam cup, cardboard cover
thermometer
small piece of metal (copper, aluminum, brass, etc.)
400-mL beaker
$NH_4NO_3(s)$, $CaCl_2(s)$ anhydrous
chemistry handbook(s)

KEYED OBJECTIVES IN TEXT: 6-2, 6-5, 6-6

DISCUSSION OF EXPERIMENT

Heat causes an increase in the temperature of a subtance which makes a solid melt or a liquid boil.   The amount of heat required to raise the temperature of 1 gram of a substance 1 degree Celsius is called its specific heat. A calorie is defined as the amount of heat that causes an increase of 1 degree Celsius for 1 gram of water. By measuring the amount of water and the change in its temperature, the heat change can be calculated.

calories =   mass of water   x   temperature change x specific heat

When a substance dissolves in water, the temperature of the water will often change. If the temperature increases, heat has been given off by the dissolving process. We say that the reaction is exothermic. If the temperature of the water drops, heat has been absorbed from the water, and we say that the reaction is endothermic. The change in temperature during the solution process is utilized in first aid packs called hot packs and cold packs. A pack contains water, and a substance isolated in a vial. When the pack is hit or squeezed, the vial breaks, and the substance dissolves in the water causing a temperature change. If the change is exothermic, the pack becomes hot; if the change is endothermic, the pack becomes cold.

Two typical substance used in hot packs and cold packs are ammonium nitrate, NH_4NO_3 , and anhydrous calcium chloride, $CaCl_2$. You will determine which substance is used in a hot pack and which is used in a cold pack. From the temperature change, the mass of the substance, and the mass of the water, you will be able to calculate the heat of solution in kcal/mole of substance.

1 kilocalorie (kcal) =   1000 calories (cal)

EXPERIMENT 17   ENERGY MEASUREMENT

LABORATORY ACTIVITIES

PART I: SPECIFIC HEAT

A.   COLLECTING DATA

A-1   Obtain a small piece of metal or other substance suggested by
      your instructor. Be sure you can pick up the object with your
      tongs since it will be heated later in the experiment. If
      not, tie a piece of thread to the object that is long enough
      to extend over the edge of the beaker while heating. Weigh
      the object and record its mass (g).

A-2   Place 250-300 mL of water in a 400 mL beaker and bring to a
      vigorous boil. Using tongs or the attached thread, place the
      metal object in the boiling water. Allow the object to remain
      in the boiling water for at least 3 minutes. Record the
      temperature of the boiling water.

A-3   While the metal is heating in the boiling water, prepare the
      calorimeter. Weigh the calorimetry cup. Record its mass. Add
      50 g water (or 100 g water for larger metal pieces). Record
      the combined mass of the calorimetry cup and water.

A-4   When you are about ready to transfer the heated metal object,
      determine the temperature of the water in the calorimeter.
      See Figure 17-1. Record the temperature of the water in the
      laboratory record.

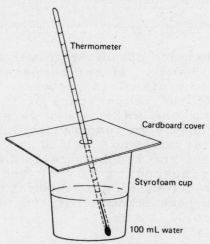

Figure 17-1   Apparatus for calorimetry.

A-5   Using tongs or the attached thread, remove the heated metal
      object from the boiling water bath, and quickly transfer it to
      the calorimeter. Replace the cover.   Stir gently as you
      watch the temperature. When the temperature remains constant
      or begins to drop, record the highest temperature reached.

126

EXPERIMENT 17   ENERGY MEASUREMENT

B. CALCULATION OF SPECIFIC HEAT

When the heated metal object is placed in the calorimeter, the metal
loses heat to the water. The change in temperature of the water can
be used to calculate this heat loss.

B-1 Calculate the temperature change (ΔT) for the water.

B-2 Calculate the calories gained by the 50 g (or 100 g) of water
     in the calorimeter using the specific heat of water which is
     1.00 cal/g$^{\circ}$C.

$$\text{calories} = \text{mass of water} \times (\Delta T) \times \frac{1.00 \text{ cal}}{(g)(^{\circ}C)}$$

B-3 Since we assume that the heat gained by the water is equal to
     the heat lost by the metal object, we can state the calories
     lost by the metal.

$$\text{heat loss(metal)} = \text{heat gain (water)}$$

B-4 Calculate the temperature change for the metal. Subtract the
     final temperature of the metal (final temperature of the water
     in the calorimeter) from the temperature of the boiling water,
     assuming that the metal object reached these same temperatures.

B-5 Calculate the specific heat of the metal.

$$\text{specific heat} = \frac{\text{calories lost by metal}}{(\text{mass of metal})(\Delta T \text{ of metal})}$$

B-6 If possible, obtain a chemistry handbook and look up the
     actual specific heat for the metal. Record.

PART II: HEAT OF SOLUTION

C. DATA

C-1 Weigh a calorimetry cup. Record.   Add 100 g of water to the
     cup and reweigh. Record the combined mass.

C-2 Weigh out 8.00 ammonium nitrate which is 0.10 mol. Record the
     mass in grams.

C-3 Record the temperature of the water in the calorimetry cup.
     Add the ammonium nitrate to the water and replace the cover.
     Stir gently to mix the solution. Observe the temperature
     until it reaches a constant value or begins to return to room
     temperature. Record the final temperature of the solution.

Empty the contents of the styrofoam cup, and rinse. Repeat the above
procedure using 11.10 g (0.10 mol) calcium chloride (anhydrous).

EXPERIMENT 17   ENERGY MEASUREMENT

D.   <u>CALCULATIONS</u>

D-1   Calculate the mass of water added to the calorimetry cup.

D-2   Calculate the number of moles of ammonium nitrate (or calcium chloride).

D-3   Calculate the change in temperature (ΔT) for each sample.

D-4   Calculate the calories for the heat of solution.

$$\text{calories} = \text{mass of water} \times (\Delta T) \times \frac{1.00 \text{ cal}}{g \ ^{\circ}C}$$

D-5   Calculate the number of kilocalories.

D-6   Calculate the heat of solution by dividing the kilocalories by the moles of compound present.

$$\text{heat of solution} = \frac{\text{kcal}}{\text{moles}} = \text{kcal/mole}$$

D-7   State whether the reaction was exothermic or endothermic, and if the compound in the reaction would typically be used in a hot pack or a cold pack.

D-8   Using a handbook, record the actual values for the heat of solution (kcal/mole) of these two compounds.

128

EXPERIMENT 17   ENERGY MEASUREMENT

LABORATORY RECORD

NAME_____

DATE_____

PART I: SPECIFIC HEAT

SECTION_____

A.   COLLECTING DATA

A-1   Type of metal   _____

Mass of metal   _____

A-2   Temperature of boiling water   _____

A-3   Mass of calorimetry cup and water_____

Mass of calorimetry cup   _____

A-4   Initial Temperature of water   _____

A-5   Final Temperature of water   _____

B.   CALCULATION OF SPECIFIC HEAT

B-1   Temperature change of water   _____

B-2   Calories gained by water   _____
      Calculations:

B-3   Calories lost by metal   _____

B-4   Temperature change by metal   _____
      Calculations:

B-5   Specific heat of metal   _____
      Calculations:

B-6   Specific heat(handbook value)   _____

EXPERIMENT 17   ENERGY MEASUREMENT

LABORATORY RECORD   NAME_____

PART II: HEAT OF SOLUTION

C. <u>DATA</u> NH_4NO_3 $CaCl_2$

C-1   Mass of calorimetry cup   _____   _____

      Mass of calorimetry cup
      and water   _____   _____

C-2   Mass of salt   _____   _____

C-3   Initial temperature of water _____   _____

      Final temperature of water   _____   _____

D.   <u>CALCULATIONS</u>

D-1   Mass of water   _____   _____

D-2   Moles of salt   _____   _____

D-3   Temperature change (ΔT) _____   _____

D-4   Calories   _____   _____
      Calculations:

D-5   Kilocalories (kcal)   _____   _____

D-6   Heat of solution(kcal/mole) _____   _____
      Calculations:

D-7   Exothermic or endothermic   _____   _____

      Hot pack or cold pack   _____   _____

D-8   Heat of solution(handbook)   _____   _____

130

EXPERIMENT 17   ENERGY MEASUREMENT

QUESTIONS AND PROBLEMS

1.   Write the standard expression and value for the specific heat
     of water(liquid).

2.   Why do materials such as copper, silver or tin tend to get hot
     very quickly when heated?

3.   Water has one of the largest specific heats of any substance.
     Why is this important for the human body?

4.   How many calories of heat are required to raise the
     temperature of 225 g of water from $42^\circ C$ to $75^\circ C$?

5.   An 80.0 g piece of metal is transferred from a hot water bath
     at $100^\circ C$ to a calorimeter containing 200.0 g of water at $22^\circ C$.
     The final temperature in the calorimeter is $28^\circ C$. What is the
     specific heat of the metal?

6.   Using your experimental value for heat of solution and the
     true value from the handbook, calculation a percentage error
     for the heat of solution for ammonium nitrate.

     %error   =   difference between true and experimental   x 100
                                      true

131

EXPERIMENT 18   HEATS OF FUSION AND VAPORIZATION; CALORIC VALUES OF FOOD

PURPOSE

1. Experimentally determine the heats of fusion and vaporization for water.
2. Use calorimetry to determine the caloric value of a food.

MATERIALS

calorimeter setup or a styrofoam cup and cardboard cover
thermometer
ice cubes
stopper(1-hole)fitted with glass tubing to fit an Erlenmeyer flask
rubber tubing
beaker (250 mL)
water
aluminum can
wire screen
food such as chips, nuts, sugar cube, or crackers

KEYED OBJECTIVES IN TEXT: 6-5, 6-6

DISCUSSION OF EXPERIMENT

The amount of heat required to melt a substance or lost when a substance freezes is called the heat of fusion. The heat of fusion for water is 80 cal/g. There is no temperature indicated since the heat of fusion occurs at the melting(freezing) point and is constant during the change of state. When a substance changes from liquid to gas at the boiling point, the heat or energy required is called the heat of vaporization. To change 1 g of liquid water into gaseous water, 540 cal are required. The heat of vaporization for water is 540 cal/g. In the first two parts of this experiment, you will experimentally determine the heat of fusion and the heat of vaporization for water.

We can also calculate the energy of foods using the specific heat of water (1 cal/g^{o}C). The foods are usually broken down into three major categories and the energy determined:

food category	kcal/g
carbohydrate	4
fat	9
protein	4

The caloric value of a food can be experimentally determined by burning a food and using the energy to heat a quantity of water. An aluminum can is used to increase the heat conductivity. By calculating the calories absorbed by the water, the number of calories provided per gram of the food can be calculated.

133

EXPERIMENT 18   HEATS OF FUSION AND VAPORIZATION: CALORIC VALUES

LABORATORY ACTIVITIES

PART I: HEAT OF FUSION

A.   <u>EXPERIMENTAL</u> <u>DATE</u>

A-1   Weigh the calorimetry cup and record its mass. Add 100 g of water to the cup and reweigh. Record the total mass in grams.

A-2   Measure the temperature of the water in the calorimeter. Record.

A-3   Place a medium-sized ice cube in the calorimeter. Stir while you watch the temperature drop. When the water temperature drops to 3 or 4°C or lower, remove any unmelted ice cube. Record the final temperature.

A-4   Reweigh the calorimetry cup and water.   It will weigh more due to the melted ice.

B.   <u>CALCULATIONS</u> <u>FOR</u> <u>THE</u> <u>HEAT</u> <u>OF</u> <u>FUSION</u>

B-1   Calculate the mass of water initially added to the calorimeter.

B-2   Calculate the temperature change for the water.

B-3   Calculate the calories lost by the water.

calories   =   mass of water   x ΔT x   1.00 cal/g°C

B-4   State the number of calories absorbed by the ice to melt. (This heat gain will be equal to the heat loss by the water).

B-5   Determine the amount of ice added to the calorimeter. Subtract the mass of the calorimetry cup and water (A-1) from the mass of the cup, water and melted ice (A-4).

B-6   Calculate the heat of fusion for ice by dividing the calories required to melt the ice (B-4) by the mass of the ice (B-5).

heat of fusion: $\dfrac{\text{calories to melt ice}}{\text{mass of ice}}$ =   cal/g

134

PART II: HEAT OF VAPORIZATION (DEMONSTRATION OPTIONAL)

C.   COLLECTING DATA

C-1   Obtain a 1-hole stopped fitted with a short length of glass tubing that fits snugly in a 250-mL Erlenmeyer flask. Attach a piece of rubber tubing to the glass tubing in the stopper. Fill the flask with about 150 mL water. Set on a wire screen on a ringstand, hold in place with the utility clamp if necessary, and bring the water to a boil.   Steam will begin to flow out of the rubber tubing.   See Figure 18-1.

CAREFUL!   THIS MUST BE DONE UNDER A TEACHER'S SUPERVISION. STEAM HAS A HIGH HEAT OF VAPORIZATION AND CAUSES SEVERE BURNS.

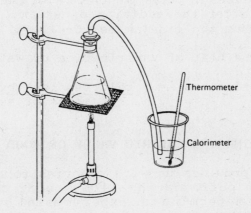

Figure 18-1.   Apparatus for measuring the heat of vaporization.

While the water in the flask is heating, weigh a calorimetry cup. Record its mass.   Place 100 g of water in the cup and reweigh. Record the total mass.

C-2   Determine the temperature of the water in the calorimeter setup.   Record.

C-3   When the water in the Erlenmeyer flask is boiling, carefully place the end of the rubber tubing in the water of the calorimeter.   CAREFUL: HOT STEAM WILL BE COMING OUT OF THE TUBING.   Cover as well as possible and read the temperature of the water as the steam condenses inside the calorimeter. When the final temperature reaches 70 or 80°C, remove the tubing and turn off the burner under the boiling water. Record the final temperature of the hot water in the calorimeter.

C-4   Weigh the cup and water contents.   There will be more water due to the condensation of the steam.   Record the total mass.

D. CALCULATIONS FOR HEAT OF VAPORIZATION

D-1   Calculate the mass of water in the calorimeter.

D-2   Calculate the temperature change for the water.

D-3   Calculate the calories gained by the water in the calorimeter.

calories = mass of water   x ΔT x   1.00 cal/g^{o}C

D-4   State the number of calories lost by the steam upon condensation. This is equal to the calories gained by the water.

D-5   Calculate the amount of steam that condensed by calculating the difference in mass of the calorimeter and its contents before and after the addition of steam. Subtract the initial mass(C-1) from the final mass (C-4).

D-6 Calculate the heat of vaporization of water.

$$\text{Heat of vaporization} = \frac{\text{calories lost by steam}}{\text{mass of steam}} = \text{cal/g}$$

PART III: MEASURING THE CALORIC VALUE OF FOOD

The food we eat provides us with calories for energy. Using a food such as a cheese puff, marshmallow, sugar cube, pretzel, nut or other, you will determine an experimental value for the caloric value(kcal/g) of that food item.

E.   DATA

E-1   Determine the mass of an aluminum can. Record. Add 50 or 100 g of water to the can. Record the combined mass.   Prepare the setup as seen in Figure 18-2. Place the aluminum can on an iron ring. A second iron ring should be placed a short distance below the aluminum can. Suspend a thermometer from a clamp so the bulb is below the water level in the aluminum can.

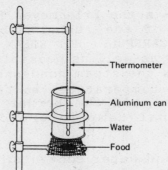

Figure 18-2 Apparatus for determination of caloric value of food.

E-2   Record the temperature of the water(initial) in the can.

E-3   Obtain the food sample you are going to use in the combustion reaction.   State the type of food.   Weigh the sample and record.

E-4   Use a match or the flame from the Bunsen burner to ignite the food sample.   Remove the flame immediately and let the food sample burn. Weigh any remainder of the food sample and record.

F. CALCULATIONS

F-1   Calculate the mass of water in the aluminum can.

F-2   Calculate the temperature change of the water before and after the combustion reaction.

F-3   Calculate the calories of heat absorbed by the water.

   calories   =   mass of water   x ΔT x   1.00 cal/$^{\circ}$C

F-4   Calculate the calories absorbed by the aluminum can assuming the same temperature change.   Specific heat of Al = 0.22 cal/g$^{\circ}$C

   calories   =   mass Al can   x ΔT x   0.22 cal/g$^{\circ}$C

F-5   Total the calories in F-3 and F-4 and express in kilocalorie units.

F-6   Calculate the mass of food sample that reacted.   Subtract the mass of the remaining food after combustion from the initial mass of the food sample.   Record.

F-7   Determine the caloric value of the food by dividing the kilocalories of heat by the reacting mass of the food sample.

   caloric value   = $\dfrac{kcal}{g\ food\ reacted}$

137

EXPERIMENT 18   HEATS OF FUSION AND VAPORIZATION: CALORIC VALUE
                OF FOOD

LABORATORY   RECORD

NAME_____
DATE_____
SECTION_____

A.   <u>EXPERIMENTAL</u> <u>DATE</u> <u>FOR</u> <u>HEAT</u> <u>OF</u> <u>FUSION</u>

A-1   calorimetry cup and water   _____

      calorimetry cup   _____

A-2   initial temperature of water   _____

A-3   final temperature of water   _____

A-4   calorimetry cup, water and melted ice   _____

B.   <u>CALCULATIONS</u> <u>FOR</u> <u>THE</u> <u>HEAT</u> <u>OF</u> <u>FUSION</u>

B-1   mass of water   _____

B-2   temperature change for water   _____

B-3   calories lost by water
      calculation: _____

B-4   calories to melt ice   _____

B-5   mass of ice   _____

B-6   Heat of fusion
      calculation: _____

NAME_____

C. COLLECTING DATA

C-1   mass of calorimetry cup and water   _____

      mass of calorimetry cup   _____

C-2   temperature of water(initial)   _____

C-3   temperature of hot water(final)   _____

C-4   mass of calorimetry cup, water and
      condensed steam   _____

D. CALCULATIONS FOR HEAT OF VAPORIZATION

D-1   mass of water   _____

D-2   temperature change for heated water   _____

D-3   calories gained by water   _____
      calculation:

D-4   calories lost by steam   _____

D-5   mass of steam   _____

D-6   Heat of vaporization   _____
      calculation:

PART III: MEASURING THE CALORIC VALUE OF FOOD

E.   DATA

E-1   Mass of Al can and water   _____

      Mass of Al can   _____

E-2   Temperature of water(initial)   _____

E-3   Type of food   _____

      Mass of food   _____

E-4   Mass of food after heating   _____

F.   CALCULATIONS

F-1   Mass of water   _____

F-2   Temperature change of water   _____

F-3   calories gained by water
      calculation: _____

F-4   calories gained by Al can
      calculation: _____

F-5   total calories gained   _____

      total in kilocalories   _____

F-6   Mass of food sample reacted   _____

F-7   Caloric value
      calculation: _____

EXPERIMENT   18   HEATS OF FUSION AND VAPORIZATION; CALORIC VALUES OF
                   FOOD

QUESTIONS AND PROBLEMS   NAME_____

1. How many calories are required to melt 25 g of ice at 0°C?

2. Draw a heating curve for water that begins with ice at 0°C
    and goes to liquid water at 60°C. Label each part.

3. Observe the heating curve in question 2 and calculate the
    kilocalories needed to convert 75 g of ice at 0°C to liquid
    water at 60°C?

4. Calculate the number of calories released when 5 g of steam
    condense.

5. Compare the amount of heat(kcal) released by 150 g of water at
    100°C hitting the skin and cooling to body temperature(37°C) to
    150 g of steam that hits the skin and cools to body
    temperature.   Why are steam burns so severe?

EXPERIMENT 19   BOYLE'S LAW

PURPOSE

1.   Observe the effects of a pressure change upon the volume of a
     gas.
2.   Graph the relationship between pressure and volume.
3.   Describe the relationship between the pressure and volume of a
     gas.
4.   Calculate the constant PxV for the experiment.

MATERIALS

Mercury tube apparatus

KEYED OBJECTIVES IN TEXT: 7-1, 7-2, 7-3, 7-4

DISCUSSION OF EXPERIMENT

To test the relationship between the pressure and the volume a gas,
your instructor will use or demonstrate the use of a mercury tube
apparatus.   The pressure is changed by moving the reservoir of
mercury to higher or lower levels.   The new volume of air in the
closed tube is measured each time a new pressure is attained.   The
variables of temperature and moles do not change during the
experiment;   they are constant during the experiment.

Preparation of the Mercury Tube Apparatus

     Before you proceed, make sure the mercury levels in both glass
tubes are about equal, then open the stopcock which allows air to
enter.   The levels of mercury will become equal. The pressure of
the air in both glass tubes is equal to the atmospheric pressure.
See Figure 19-1.

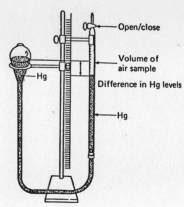

Figure 19-1   Apparatus used to measure the relationship of
               pressure and volume

Closing the stopcock causes a certain volume of air to become trapped in the closed tube. Raising or lowering the mercury reservoir changes the pressure on the air in the closed tube. As a result, the volume in the closed tube also changes. Each time the mercury reservoir is moved, a new pressure and volume are measured.

When the level of mercury in the closed tube is equal to the mercury level in the open tube, the pressure of the air sample is equal to atmospheric pressure. When the level of mercury in the closed tube is lower than the level of mercury in the open tube, the pressure of the air sample is greater than atmospheric pressure.

$$P_{closed} = P_{atm} + P_{difference}$$

When the mercury level in the closed tube is higher than the mercury level in the open tube, the pressure of the air sample is lower than atmospheric pressure.

$$P_{closed} = P_{atm} - P_{difference}$$

EXPERIMENT 19   BOYLE'S LAW

LABORATORY ACTIVITIES(OPTIONAL TEACHER DEMONSTRATION)

A. PRESSURE AND VOLUME OF A GAS

A-1 Read a barometer and record the barometric pressure in mmHg.

A-2 Open the stopcock of the mercury apparatus and allow the levels of mercury to become equal. Then close the stopcock to trap the air sample inside the closed tube. This will be the air sample(1) for the experiment. Record the volume of the air sample. Read the levels of mercury in both the closed and open tubes and record their values in mm Hg.

A-3 Raise or lower the mercury reservoir to give a new pressure and volume. Record the volume of each new air sample and the levels of mercury in the closed and open tubes.

B.   CALCULATIONS

B-1 For each sample, calculate the pressure difference in mmHg for the mercury levels of the closed and open tube. If the mercury level in the closed tube is lower than the open tube, place a positive (+) sign in front of the difference. If the mercury level in the closed tube is higher than the open tube, place negative(-) sign in front of the pressure difference.

B-2 Add or subtract the pressure difference calculated in B-1 to the atmospheric pressure. This will be the pressure of the air sample in the closed tube.

$$P_{air\ sample} = P_{atm}\ (+\ OR-)\ P_{difference}$$

B-3 Boyle's Law indicates that P x V should be a constant value for a sample of gas at constant temperature and moles. We can test our data by multiplying the pressure and volume values. For each air sample, multiply the pressure of the sample (B-2) and the volume. Round off to the correct number of significant figures.

B-4 Using graph paper from the back of this lab book, graph the data obtained in this experiment. Place the pressure(mm Hg)on the vertical axis and the volume(ml) on the horizontal axis. Use the full area of the graph paper by adjusting the lowest and highest values to fit. The lowest value should be near the lowest values obtained in the data. (You do not need to start at zero.) Be sure to mark the units of pressure and volume in equal intervals. Draw a smooth line (slight curve) through the points obtained from the data.

B-5 Use the graph to discuss the meaning of Boyle's Law.

EXPERIMENT 19 BOYLE'S LAW

LABORATORY RECORD

NAME ____________________

SECTION ____________________

DATE ____________________

BAROMETRIC PRESSURE ____________

A. DATA

B. CALCULATIONS

READINGS	VOLUME (mL)	MERCURY LEVEL closed	MERCURY LEVEL open	PRESSURE DIFFERENCE	PRESSURE OF AIR SAMPLE	P X V
1						
2						
3						
4						
5						

QUESTIONS AND PROBLEMS

1. The behavior of a gas according to Boyle's Law requires that
    the temperature and number of moles of a gas be kept constant.
    How did this experiment keep the temperature and the moles of
    gas constant?

2. Using the graph you prepared in B-4, explain the relationship
    between pressure and volume according to Boyles Law.

3. Why is the product P x V constant (or very close) for the air
    sample measured at different pressures and volumes?

4. Indicate what should happen to the pressure or volume in the
    following examples when T and n are constant:

               Pressure   Volume

               increases   _____

               _____   increases

               decreases   _____

148

EXPERIMENT 20   CHARLES' LAW

PURPOSE

1. Observe the effect of changes in temperature upon the volume of a gas.
2. Graph the data for volume and temperature of a gas.
3. State a relationship between the temperature and volume of a gas.

MATERIALS

The following setup may be made available by your instructor:
   Erlenmeyer flask (125 or 250 mL)
   one-hole rubber stopper fitted with a short piece of
      glass(polished)

_____

  400-mL beaker (larger for a 250-mL Erlenmeyer flask)
  graduated cylinder
  water trough or large container or bucket
  thermometer
  ice

KEYED OBJECTIVES IN TEXT: 7-5, 7-6

DISCUSSION OF EXPERIMENT

The way that the volume of a gas changes with a change in Kelvin temperature will be illustrated in this experiment. The volume of air contained in an Erlenmeyer flask will serve as the gas sample. A temperature of $100^{\circ}C(373K)$ can be obtained by placing the flask in a boiling water bath. The volume of the gas at the boiling point of water will be the full volume of the flask. The temperature of the air contained in the flask is lowered by submerging the flask into water baths of different temperatures around $40-50^{\circ}C$ and $0^{\circ}C$. These measurements will be taken while the variables of pressure and number of moles of gas are held constant.

The volume of a gas changes directly with the Kelvin temperature as long as the pressure and number of moles are held constant. This can be expressed mathematically as

$$V \propto T$$

or

$$\frac{V}{T} = constant \qquad \frac{V_1}{T_1} = \frac{V_2}{T_2}$$

EXPERIMENT 20   CHARLES' LAW

LABORATORY ACTIVITIES

A.   <u>TEMPERATURE</u> <u>AND</u> <u>VOLUME</u> <u>OF</u> <u>A</u> <u>GAS</u>

A-1 Place the rubber stopper and glass tube in the Erlenmeyer
    flask.   Attach a buret clamp to the neck of the flask and
    place in a 400-mL beaker. Add water up to the neck of the
    flask. Allow enough space for the water to boil so it does not
    splatter or boil over. See Figure 20-1.   Set the beaker with
    the flask and water on an iron ring covered with a wire screen
    and bring the water to a boil.   Continue to boil(gently) for
    5 more minutes. Record the temperature of the boiling water.

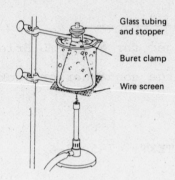

Glass tubing
and stopper

Buret clamp

Wire screen

Figure 20-1   Setup for flask and air in a boiling water bath.

A-2 <u>Water</u> <u>Bath</u>: Obtain a water trough or a deep container(bucket)
    and fill about 2/3 full with tap water. Some warmer water may
    be added to raise the temperature of the water bath. Use a
    temperature of about 30-40°C.

    Turn off the burner. Prepare to remove the flask. Place
    your finger tightly over the end of the glass tubing, and lift
    the flask out the hot-water using the buret clamp as a handle.
    BE CAREFUL WHILE WORKING WITH A BOILING WATER BATH.   While
    holding the <u>stopper</u> <u>end</u> <u>downward</u>, immerse the flask into the
    water trough or container of water. See Figure 20-2.

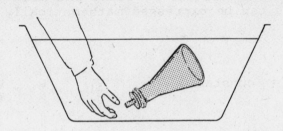

Figure 20-2   Placing the flask in a water trough.

150

The reduction in temperature will cause a decrease in the volume of the air in the flask, and you will see water enter the flask. Let the flask remain completely submerged in the water for 5 minutes or more. KEEP THE FLASK INVERTED THE ENTIRE TIME. THE AIR INSIDE MUST REMAIN TRAPPED. Record the temperature of the water in the water trough.

A-3 To adjust the pressure inside the flask with the atmospheric pressure, it is necessary to equalize the levels of water inside and outside the flask. **Keeping the flask inverted(upside down)**, slowly move the flask to a height in the water until the water level inside the flask matches the water level outside. See Figure 20-3. When they are equal, tightly cover the glass tubing with your finger under the water, and then remove the flask from the water and set it upright on your desk. It is important to retain the water than entered the flask during the cooling.

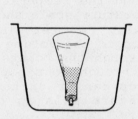 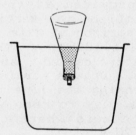

Figure 20-3   Equalizing the water levels inside and outside the inverted flask.

Use a graduated cylinder to measure the volume of water found in the flask. Record.

A-4 Dry the flask, replace the stopper and glass tubing, and heat in a boiling water bath as you did in step A-1. While the water is heating, prepare an ice-water mixture in the water trough or container. After the flask has heated in the boiling water for at least 5 minutes, place your finger over the tubing, invert the flask and submerge it in the cold ice-water mixture. Allow the flask to remain submerged for at least 5 minutes. Record the temperature of the cold water.

A-5 Equalize the water levels, cover the glass tubing, remove the flask and contained water from the trough and set on your desk. Use a graduated cylinder to measure the volume of water that entered the flask. Record.

A-6 Determine the full volume of the Erlenmeyer flask by filling it with water up to the level of the rubber stopper. Measure the volume of water with a graduated cylinder.

B.   CALCULATIONS

B-1   Determine the volume of air in the flask as V_{boil}. This is
      equal to the full volume of water that filled the flask in
      step A-6.

B-2   Determine the volume of air in the flask in the warm and cold
      water by subtracting the volume of water that entered the
      flask from the full volume of the flask.

      $$V_{air} \quad = \quad V_{full} \quad - \quad V_{water}$$

B-3   Convert the temperatures to kelvins.

B-4   Calculate the V/T constants.

B-5   Graph the volume and temperature of a gas on the graph
      paper provided at the back of this lab book. Place the
      volume(mL) on the vertical axis and the temperature(^{O}C) scale
      on the horizontal axis. The temperature scale should run from
      -400°C to 100°C.   Theoretically,   matter would reach a
      temperature called **absolute zero** if the volume were reduced
      to zero. Draw a straight line through the points on the graph.
      By extending the line that goes through your experiment data
      all the way to zero volume , you can predict the value of
      absolute zero where the temperature axis is crossed. See
      Figure 20-4.

Graph of Volume vs Temperature (P,N constant)

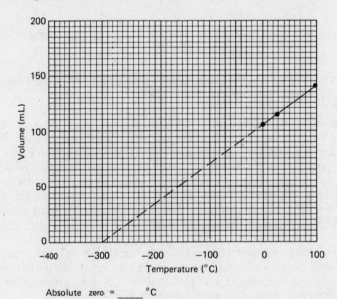

Absolute zero = _____ ^{O}C

Figure 20-4   Example of graph for volume and temperature.

EXPERIMENT 20   CHARLES' LAW   NAME_____

LABORATORY RECORD   DATE_____

                                        SECTION_____

A.   TEMPERATURE AND VOLUME DATA

READINGS   IN

	BOILING WATER	WARM WATER	COLD WATER
Temperature(^{o}C)			
Volume of H_2O entering flask	0 ml		
Full Volume of flask			
B. CALCULATIONS			
Volume of air			
Temperature (K)			
V/T			

Results of graph

     Temperature at zero volume _____

153

EXPERIMENT 20   CHARLES' LAW   NAME_____
                                             DATE_____
                                             SECTION_____

QUESTIONS AND PROBLEMS

1.   Using the graph you prepared in B-5, discuss the relationship
     between volume and temperature according to Charles' Law.

2.   How does your predicted value for absolute temperature compare
     with the actual value $-273^{O}C$?

3.   A gas with a volume of 525 mL at a temperature of $-50^{O}C$ is
     heated to $200^{O}C$. What is the new volume of the gas if pressure
     and number of moles are held constant?

4.   A gas has a volume of 2.8 L at a temperature of $10^{O}C$ What
     temperature is needed to expand the volume to 44 L? (P,n
     constant)

EXPERIMENT 21   PARTIAL PRESSURES OF OXYGEN
                       AND CARBON DIOXIDE IN AIR

PURPOSE

1. Measure the percentage of oxygen in the air.
2. Calculate the partial pressures of oxygen and nitrogen in the air.
3. Measure the partial pressure of CO_2 in the atmosphere and in expired air.

MATERIALS

PART I: iron filings
          250-mL beaker
          large test tube
          graduated cylinder
          label

PART II: 250-mL Erlenmeyer flask
          shell vials
          6 M NaOH
          mineral oil
          two-holed rubber stopper with two short pieces of
              glass tubing inserted
          rubber tubing
          pinch clamps
          straws to fit into end of rubber tubing
          100-mL beaker
          food coloring(optional)
          meter stick

KEYED OBJECTIVE IN TEXT: 7-9

DISCUSSION OF EXPERIMENT

The two major gases in the air are oxygen (O_2) and nitrogen (N_2). You will be able to determine the percentage of oxygen in the air by removing the oxygen from the a sample of air and calcuting the corresponding volume change. As the oxygen is removed from the air sample in a large test tube, its volume in the initial air sample is replaced by water. The volume replaced is equal to the amount of oxygen in that sample. The percentage of oxygen can then be calculated. If the atmospheric pressure in mm Hg is also known, you can calculate the partial pressures of both oxygen and nitrogen in the air.

When the body metabolizes nutrients, one of the end products is carbon dioxide, CO_2. Certain levels of carbon dioxide trigger breathing mechanisms and maintain the correct pH of the blood. Accumulation of carbon dioxide above these levels can result in respiratory and metabolic dysfunction and possible death. The body eliminates the majority of its carbon dioxide by way of the lungs. Partial pressures are those individual pressures exerted by each of the gases that make up the total atmospheric pressure.

EXPERIMENT 21   PARTIAL PRESSURES

In this experiment, you will compare the partial pressures of carbon dioxide in the atmosphere and in expired(exhaled) air. Mineral oil is placed on top of the sodium hydroxide to prevent its reaction with carbon dioxide until the whole system is closed.   By tipping over the vial inside the flask, the sodium hydroxide is released and reacts with the carbon dioxide in the air samples.

$$CO_2(g) \quad + \quad NaOH \quad \longrightarrow \quad NaHCO_3$$

The other major gases in the atmosphere, oxygen and nitrogen, do not react with sodium hydroxide, and they will continue to exert their respective partial pressures.   The difference between the atmospheric pressure and the pressure of the remaining gases is used to calcalate the partial pressure of CO_2

LABORATORY ACTIVITIES

PART I: PARTIAL PRESSURES OF OXYGEN AND NITROGEN IN AIR

A.   PREPARATION AND COLLECTING DATA

A-1   Completely fill a large test tube with water.   Measure the volume of the water contained in the test tube by emptying the water into a graduated cylinder.   This volume corresponds to the initial volume of air.   Record the volume of the test tube.

A-2   Obtain a small scoop of iron filings.   Place the filings in the empty, but still moist test tube.   If the test tube is too dry, add some water to moisten the sides.   Shake the iron filings about the test tube.   They should adhere to the sides. Shake out the excess filings and discard.

Fill the 250-mL beaker about one-half full of water. Invert the test tube containing the iron filings and set it in the water.   Attach a test tube holder to support the inverted test tube.   Carefully place the beaker and inverted test tube in your drawer.   The experiment will not be finished until the next laboratory period.   See Figure 21-1.

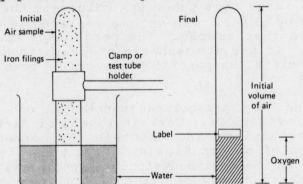

Figure 21-1   Apparatus for determination of the partial pressures of oxygen and nitrogen in the air.

156

At the next laboratory period, you should find that water entered the test tube to replace the oxygen in the air sample that reacted with the iron filings. Using a label or marker pen, mark the water level inside the test tube. Remove the test tube.

Fill the test tube with water up to the line you have marked. Empty this amount of water into a graduated cylinder. This volume of water represents the portion of air, mostly nitrogen, that remained in the test tube after the reaction. Record the volume of the nitrogen.

A-3   Read a barometer and record the barometric pressure in mmHg.

B.   CALCULATIONS OF THE PARTIAL PRESSURES OF OXYGEN AND NITROGEN

B-1   Subtract the volume measured in step A-2 from the total volume of the test tube measured in step A-1. The difference in the two volumes is equal to the volume of oxygen in the air sample. Record.

B-2   Calculate the percentage of oxygen in the air by dividing the volume of oxygen by the total volume of the air sample. Repeat the calculation for the percentage of nitrogen in the air sample. Record.

$$\frac{\text{volume(oxygen)}}{\text{volume (air sample)}} \quad \text{x } 100 \quad = \quad \text{\%oxygen in air}$$

B-3   Calculate the partial pressure of oxygen by multiplying the percent oxygen by the atmospheric pressure. Record. Repeat the calculation to determine the partial pressure of nitrogen.

$$\frac{\text{volume}(O_2)}{\text{volume (air)}} \quad \text{x} \quad \text{atmospheric pressure} \quad = \quad \text{partial pressure } O_2$$

PART II: CARBON DIOXIDE IN THE ATMOSPHERE AND EXHALED AIR

C.   PREPARATION AND COLLECTING DATA

C-1   Carefully lower a shell vial into the Erlenmeyer flask so that
      it sits upright. Set a funnel in the shell vial in the flask.
      Pour 6 M NaOH into the vial until the vial is about 3/4 full.
      Pour a thin layer(1mm) of mineral oil on top of the NaOH.
      Remove the funnel.   See Figure 21-2.

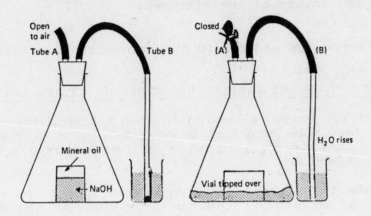

Figure 21-2 Apparatus for carbon dioxide determination.

      Place the two-holed stopper with its glass tubing in the
flask.   Attach rubber tubing to each of the pieces of glass
tubing. Attach a length of glass tubing (60-75 cm) to one of
the pieces of rubber tubing (tube B).   Place the free end of
the glass tubing in a beaker that contains water and a few
drops of food coloring (optional). With the rubber tubing
open, the flask is full of air at atmospheric pressure. Close
the system by attaching a pinch clamp to tube A.

Gently tilt the flask until the shell vial tips over and the
NaOH solution spills out into the bottom of the flask. Swirl
or shake the flask gently. (Hold the flask around the top
with your fingers. Try not to let your hands warm the flask
since the increase in temperature will change the pressure.)
Make sure that tube B remains under the water level in the
beaker. Continue to swirl the flask while you observe changes
in the water level in tube B. Wait a few minutes, then
swirl again to make sure that there is no further change in
the water level. Using a meter stick, measure the distance
(mm) between the water level in the beaker, and the water
level in tube B. Record.

$$CO_2 \quad + \quad NaOH \quad \longrightarrow \quad NaHCO_3$$

C-2   The change in water levels is related to the amount of carbon dioxide in the air that reacted with the NaOH solution. Since 13.6 mm H_2O exerts the same pressure as 1 mm Hg, we can convert the height of a column of H_2O to a corresponding height of a column of Hg. Divide the mm water by the specific gravity of mercury which is 13.6. This is the partial pressure of CO_2 in the air sample.

$$1 \text{ mm Hg} = 13.6 \text{ mm } H_2O \qquad \frac{\text{mm water}}{13.6} = \text{mm Hg}$$

C-3   Read a barometer and record the atmospheric pressure in mm Hg.

C-4   Calculate the percent CO_2 in the air.

$$\frac{\text{mmHg } CO_2(\text{air})}{\text{mmHg atm}} \times 100 = \% CO_2 \text{ (air)}$$

D. CARBON DIOXIDE IN EXPIRED AIR

D-1   Set up the apparatus as described in C-1. Then fit a clean straw into the tubing of tube A. With the system open, and with tube B in the beaker of water, take a breath of air, hold for a moment, and exhale completely through the straw. This will cause bubbling in the beaker of water. Cover the straw with your finger while you prepare for the next exhalation. Repeat the inhalation and exhalation 6 or 8 times. The atmospheric air in the flask is being replaced by exhaled air from the lungs. **Be careful that you do not get dizzy or lightheaded during this process. Stop if you do.**

With the flask filled with expired air, close the pinch clamp on tube A. (Tube B is still in the beaker of water.) Tip the shell vial inside the flask. Swirl and shake the flask gently and watch for the water to rise inside the glass tubing(B). There should be a dramatic rise in the water level inside the glass tube. If not, check your system for leaks. When no further changes occur, use a meter stick to measure the distance (mm water) between the water level in the beaker and the height of the water in tube B. Record.

D-2   Convert the height of the water column to mm Hg. This is the partial pressure of CO_2 in expired air.

$$\frac{\text{mm water}}{13.6} = \text{mm Hg}$$

D-3   Record the atmospheric pressure in mm Hg or use the same value you obtained in part C-3.

D-4   Calculate the percentage CO_2.

$$\frac{\text{mm Hg } CO_2 \text{ (exhaled air)}}{\text{mm Hg atmosphere}} \times 100 = \% CO_2(\text{in exhaled air})$$

PARTIAL PRESSURES OF OXYGEN AND CARBON DIOXIDE IN AIR

LABORATORY RECORD

NAME_____

DATE_____

SECTION_____

PART I: OXYGEN AND NITROGEN IN THE AIR

A. DATA

A-1 Volume of air sample

_____ mL

A-2 Volume of nitrogen

_____ mL

A-3 Atmospheric pressure

_____ mmHg

B. CALCULATIONS

B-1 Volume of oxygen
Calculations:

_____ mL

B-2 Percent oxygen
Calculations:

_____ $\%O_2$

B-3 Partial pressure of oxygen
Calculations:

_____ mmHg

Partial pressure of nitrogen
Calculations:

_____ mmHg

PART II: CARBON DIOXIDE IN ATMOSPHERIC AND EXPIRED AIR

C. CO_2 IN THE ATMOSPHERE

C-1   Height of water column   _____mm H_2O

C-2   Height of a mercury column   _____mmHg
      Calculations:

C-3   Atmospheric pressure   _____mmHg

C-4   Percent CO_2(atmosphere)   _____%CO_2
      Calculations:

D. CO_2 IN EXPIRED AIR

D-1   Height of water column   _____mm H_2O

D-2   Height of a mercury column   _____mmHg
      Calculations:

D-3   Atmospheric pressure   _____mmHg

D-4   Percent CO_2(expired air)   _____%CO_2
      Calculations:

QUESTIONS AND PROBLEMS

1.   In PART I, what would happen to your calculated value of the
      %O_2 in the atmosphere if some of the oxygen did not react
      with the iron filings?

2.   How would your values for the partial pressures of oxygen and
      nitrogen be affected if you were running this experiment at
      a higher altitude?

3.   What is the total pressure in mmHg of a sample of gas that
      contains 40 mmHg O_2, 1.20 atm N_2, and 800 torr He ?

4.   Why do the percentages of CO_2 differ in the atmosphere and in
      expired air?

EXPERIMENT 22   SOLUTIONS

PURPOSE
1. Determine the polarity of a solute by its solubility in polar and nonpolar solvents.
2. Determine the effect of particle size, stirring and temperature upon the rate of solution formation.
3. Identify an unsaturated and a saturated solution.
4. Graph the solubility curve for KNO_3.

MATERIALS

test tubes   solids: $KMnO_4(s)$, sucrose, $I_2(s)$
test tube rack   vegetable oil, NaCl(coarse and fine),
small beakers $KNO_3(s)$
                                    liquids: water, cyclohexane

KEYED OBJECTIVES IN TEXT: 8-1, 8-2, 8-3

DISCUSSION OF EXPERIMENT

A solution forms when one substance (solute) dissolves in another substance (solvent). When the attractive forces between the particles of the solute are similar to those between the solvent particles, the solute dissolves in that solvent. The solute and the solvent have similar polarities. Polar (or ionic) solutes dissolve in polar solvents, while nonpolar solutes require nonpolar solvents.

The rate of solution formation depends upon several factors: (1) The amount of surface area in contact with the solvent which depends upon the size of the particles; (2) The amount of shaking or stirring that occurs after a solute has been added to the solvent; and (3) The temperature of the solvent during solution formation.

A saturated solution contains the maximum amount of solute that will dissolve in 100 g of solvent at that temperature. Any additional solute will be seen as a solid precipiate. A solution with less solute is unsaturated and no precipitate is seen. Every solute has its own saturation level. For many solutes, this quantity increases with increase in the temperature.

EXPERIMENT 22   SOLUTIONS

LABORATORY ACTIVITIES

A.   POLARITY OF SOLUTES AND SOLVENTS

Set up four test tubes in the test tube racks. To test the solubility of solutes in water, place 3-4 mL of water, a polar solvent in each test tube.

Add a few crystals or a few drops, separately, of the solutes $KMnO_4$, I_2, sucrose, and vegetable oil. IODINE WILL BURN YOUR SKIN. USE TWEEZERS TO HANDLE.   Stir with a glass stirring rod. Indicate that the solute was soluble(S) if it dissolved, or insoluble (IN) if it did not dissolve.

Using four dry, clean test tubes repeat the experiment by placing 3-4 mL of cyclohexane, a nonpolar solvent, in each. CYCLOHEXANE IS FLAMMABLE: DO NOT PROCEED IF ANY GAS BURNERS ARE IN USE. Record your observations.

If a solute dissolves in a polar solvent, then the solute must be polar; if not, the solute must be nonpolar. From your observations, state whether each solute is polar or nonpolar.

B.   RATES OF SOLUTION

B-1 Effect of Particle Size.

Place some finely ground NaCl crystals in a test tube to a depth of about 1 cm. Fill another test tube to the same depth with some coarse NaCl crystals. Add 10 mL water to each. Begin stirring both samples with glass stirring rods. State which sample dissolves first.

B-2 Effect of Stirring.

Fill two test tubes with a small, but equal amounts of finely ground NaCl. Add 10 mL of water to each. Stir the sample in test tube 1, but not test tube 2. State which sample dissolves first.

B-3 Effect of Temperature

Obtain a few crystals of $KMnO_4$. Fill two beakers with 100 mL of water. Heat the water in one of the beakers to about $80^{\circ}C$. Place both beakers where they will not be disturbed. Add half of the crystals of $KMnO_4$ to the cold water and the other half to the hot water. Describe the appearance of the two solutions in 10 minutes.

166

C.   SATURATED SOLUTIONS

C-1 Obtain weighing paper or a small container and weigh carefully.
     Place between 2.50-3.50 g of KNO_3 on the paper or in the
     container and obtain the combined mass. Record. Calculate the
     mass of KNO_3.

          Place 10.0 mL of water in a large test tube. Add the
     solid. If all of the KNO_3 sample dissolves, place the test in
     an ice-water mixture. Stir the mixture with a thermometer and
     note the temperature at which the first crystals of solid KNO_3
     appear. That will be the temperature at which this solution
     reaches saturation. Record the temperature.

          Weigh out four more samples of KNO_3 between 2.50 g and
     3.50 g. Add, one by one, each of the weighed-out samples to
     the solution already in the large test tube. If the added
     solid does not completely dissolve at room temperature, place
     the test tube in a water bath and heat gently until all of the
     solid is dissolved. Stir with a thermometer. Let the test
     tube cool in the air while you continue to stir the solution
     with the thermometer. Note the temperature when the first
     crystals appear. This will be the temperature of saturation
     for this concentration.   Record a temperature for the
     appearance of crystals for each of the rest of the samples.

D.   CALCULATIONS OF SOLUBILITY AT DIFFERENT TEMPERATURE

D-1 Calculate the total mass of KNO_3 added to the water for each
     temperature.

D-2 The total mass of KNO_3 is the amount of KNO_3 that saturates
     10.0 ml of water at a particular temperature. To express
     solubility in terms of g solute/100 mL water, multiply the
     mass and volume by 10.

$$\frac{g\ KNO_3}{10\ mL\ water} \times \frac{10}{10} = \frac{g\ KNO_3}{100\ mL\ water} = solubility$$

D-3 Use your data to prepare a solubility curve for KNO_3. There is
     graph paper at the back of this lab book. Plot the solubility
     in g KNO_3/100 ml water on the vertical axis. Place the
     temperature(0-100°C) on the horizontal axis.

EXPERIMENT 22   SOLUTIONS   NAME_____

LABORATORY RECORD   DATE_____
                                            SECTION_____

A. POLARITY OF SOLUTES AND SOLVENTS

| | | | Solutes | |
	KMnO$_4$	I$_2$	sucrose	vegetable oil
water (polar)				
cyclohexane (nonpolar)				
Polarity of Solute				

State the general solubility rule concerning polarity of solute and solvent.

B.   **RATES OF SOLUTION**

B-1 Effect of Particle Size

     Sample that dissolved first   _____

What is the effect of particle size on the rate of solution?

B-2 Effect of Stirring

     Sample that dissolved first   _____

What is the effect of stirring upon the rate of solution?

B-3 Effect of Temperature (Observations after 10 min)

     Hot Water

     Cold Water

What is the effect of temperature upon the rate of solution?

169

EXPERIMENT 22   SOLUTIONS

C-1   D-1

KNO_3	MASS OF CONTAINER	MASS OF KNO_3 + CONTAINER	MASS KNO_3	TEMP. (°C)	TOTAL MASS KNO_3 IN 10g of H_2O	SOLUBILITY $gKNO3/100g$ H_2O
1						
2						
3						
4						
5						

QUESTIONS AND PROBLEMS

1. How does an unsaturated solution differ from a saturated
    solution?

2. According to your graph, how is the solubility of KNO_3 affected
    by an increase in temperature?

3. On your solubility curve, what is the change in solubility from
    $30^{\circ}C$ to $60^{\circ}C$?

4. At what temperature would the solubility of KNO_3 be

    50 g/100 g H_2O?

    100 g/100 g H_2O?

EXPERIMENT 23   WATER IN FOOD,   ELECTROLYTES   AND WATER PURITY

PURPOSE

1. Calculate the percent water in a fruit or vegetable.
2. Compare the conductivities of strong, weak and nonelectrolytes.
3. Observe the kinds of electrolytes and their concentrations(meq/L)
    in parenteral solutions.
4. Test a variety of water samples for water hardness.
5. Use water treatment techniques to purify water.

MATERIALS

PART I: WATER IN FOOD

fruit or vegetable
knife
drying oven
watch dish

PART II: ELECTROLYTES AND CONDUCTIVITY

conductivity apparatus   Liquids: distilled water,
IV solutions in bottles   Solutions: (1M) HCl, NaCl, NaOH
   with labels   HAc, NH_4OH, cyclohexane,
                                             sugar, glucose, ethanol
                                  Solids: NaCl

PART III: TESTING WATER HARDNESS AND SETTLING

test tubes
test tube rack
tincture of green soap
muddy water
water samples: distilled, hard, soft, tap, sea water
                mineral water, others

1% solutions: NaCl, $Al_2(SO_4)_3$, Na_2SO_4, $FeCl_3$

KEYED OBJECTIVES IN TEXT: 8-4,8-7

DISCUSSION OF EXPERIMENT

The foods we eat are composed of large quantities of water as are
the cells of our bodies which are about 60% water. We obtain water
for the body primarily by drinking fluids and eating food with a
high content of water. We can calculate the amount of water in a
food by dehydrating a piece of food and measuring the amount of
water lost.

Electrolytes are substances that produce ions in aqueous solutions. When ions are present in an aqueous solution, the light bulb of a conductivity apparatus will glow, because the ions complete the electrical circuit. A nonelectrolyte produces only molecular substances which do not carry current in an aqueous solution. The light bulb in the conductivity apparatus does not glow. Weak electrolytes produce only a few ions: the light bulb will glow weakly. We can identify substance in aqueous solutions as electrolytes, weak electrolytes, or nonelectrolytes on the basis of our observations of the conductivity apparatus.

The cells of the body are bathed both inside and outside by fluids that contain specific, differing amounts of electrolytes. The electrolytes play an important role in the maintenance of the activities in the cell. Inside the cell, the major cation is potassium, K^+, and the major anion is bicarbonate, HCO_3^-. The fluid within the cell is called the <u>intracellular fluid</u>. Outside the cell, in the <u>extracellular fluid</u>, the major cation is sodium Na^+, and the major anion is chloride, Cl^-. Other electrolytes are found in smaller quantities in both intracellular and extracellular fluids.

When there is a loss of fluid from the body or an imbalance of electrolytes, a parenteral solution (one given by means other than oral) may be administered. The type of parenteral solution used reflects the needs of the cells in the body as determined by laboratory tests of the body fluids. When giving parenteral solutions, keep in mind the effect of the electrolytes in maintaining and regulating fluid balance, muscle tone, and acid-base balance in the body. See Table 23-1. If there is an imbalance, there may also be a shift of water between plasma and tissues or a loss of water accompanied by a shift or loss of essential electrolytes.

Table 23-1.   Effects of Electrolytes Imbalance

EFFECTS

ELECTROLYTE	LOW LEVELS	HIGH LEVELS
Na^+, Cl^-	weakness, headache, diarrhea, cramping	Edema
K^+	apathy, cardiac changes	cardiac arrest, numbness
Ca^{2+}	numbness in extremities, tetany	deep bone pain
HCO_3^-	acidosis (low pH)	alkalosis (high pH)
Mg^{2+}	convulsions, disorientation	dehydration, coma

EXPERIMENT 23   WATER AND ELECTROLYTES

Water that contains the ions Ca^{2+}, Mg^{2+}, and Fe^{3+} is called **hard water.** The greater the concentration of these ions in the water, the greater is the degree of hardness. When hard water reacts with soap, the ions in the hard water and some of the soap molecules form an insoluble salt called a **soap scum.** The soap molecules tied up in the scum are not free to perform their cleaning function. More soap must be added to remove the ions and allow sufficient sudsing and cleaning.

The reaction of a soap solution with the ions in hard water can be used to compare hardness of water samples. In some hospitals, a soap solution may be used to test the water used for dialysis treatments, which must be free of these ions.

In water treatment stations, certain chemicals added to water will cause the formation of insoluble substances that sink to the bottom of the tank. The water on top is purified and can be drawn off for use.

LABORATORY ACTIVITIES

PART I: PERCENT WATER IN A FOOD

A.   DATA

A-1   Cut several thin slices of a fruit or vegetable. Record the type of food.

A-2   Weigh a watch glass. Record its mass.

A-3   Spread the food slice on the watch glass. Weigh the watch glass and food slices. Record.

A-4   Place the watch glass and food in a drying oven at 110°C for at least 1 hr. Remove the watch glass and dried (dehydrated) food. Cool. Weigh and record the total mass.

B.   **CALCULATIONS**

B-1   Calculate the mass of the food slices.

B-2   Calculate the mass of the dried food slices.

B-3   Calculate the amount of water lost through dehydration. (B1 - B2). Record the amount of water lost.

B-4   Calculate the percentage of water in the food.

$$\% \text{ water} = \frac{\text{mass of water lost (B-3)}}{\text{mass of food slices (B-1)}} \times 100\%$$

List the values for other foods tests. Look up handbook values for the percent water in the foods that were tested.

175

EXPERIMENT 23 WATER AND ELECTROYLTES

PART II: ELECTROLYTES AND CONDUCTIVITY

THIS PART OF THE EXPERIMENT MUST BE A DEMONSTRATION BY YOUR INSTRUCTOR. THE BARE ELECTRODES ARE A HAZARD. DO NOT TOUCH THE ELECTRODES WHEN THE APPARATUS IS PLUGGED IN. YOUR SKIN WILL CONDUCT AN ELECTRIC CURRECT AND CAUSE A SHOCK.

C-1 Select one of the solutions to test. With the conductivity apparatus plugged in, place the electrodes in the solution and observe the light bulb. Identify the solution as a strong electrolyte, a weak electrolyte, or a nonelectrolyte. See Figure 23-1. Record the light bulb intensity for each type of solution.

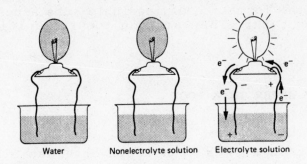

Water   Nonelectrolyte solution   Electrolyte solution

Figure 23-1   Conductivity apparatus.

C-2 On the laboratory record, indicate the kinds of particles that must be present in each solution. Strong electrolytes consist of ions in solution, weak electrolytes contain molecules and a few ions, while nonelectrolytes are substances that dissolve only as molecules.

C-3 Write the reaction that occurs when each of the solutes in the solutions tested was added to water. In equations for the weak electrolytes, use a double arrow.

EXPERIMENT 23   WATER AND ELECTROLYTES

D.   ELECTROLYTES IN BODY FLUIDS

D-1 In the classroom, there are several samples of IV bottles. Obtain one of these. Record the type of solution.

D-2 Read the label. The electrolytes present in the solution are listed on the label by their symbols. For example, "Na 47 Cl 47" means that there are 47 meq of sodium ion and 47 meq chloride ion per liter of that solution. Record these values for the cations and anions.

label	meaning
Na Cl	Na^+, 47 meq/L Cl^-, 47 meq/L
47 47	

D-3 Calculate the total number of meq of cations and record.

D-4 Calculate the total number of meq of anions and record.

D-5 Calculate the sum of the positive and negative charges and record.

Repeat the above with another IV solution.

PART III: TESTING THE HARDNESS OF WATER

E-1 Obtain 100-150 mL of a soap solution in a beaker. Set up a buret, and fill with the soap solution. Your instructor will demonstrate this procedure. Place 50 mL of the water sample you are going to test (begin with distilled water) in a 250 mL flask.   Add 1 mL of the soap solution from the buret. Stopper the flask and shake for 10 sec. With distilled water, you should see a thick layer of suds. If you don't, add another milliliter of soap solution and shake for 10 sec again. The suds that form in the distilled water sample will serve as your reference sample. Save for comparison. Reshake if necessary.

    With other water samples, continue to add soap solution until you have softened the water sample and it forms suds like the distilled water sample. Use an assortment of water samples available in the lab or from your home, pool or well. Record the number of mL required to soften each water sample.

EXPERIMENT 23   WATER IN FOOD,   ELECTROLYTES AND WATER PURITY

F.   PURIFICATION OF WATER

F-1   Set up five test tubes in a test tube rack.   To each, add 10 mL of muddy water.   Add the following chemicals to the muddy water; label each.

    (1)   5 mL water
    (2)   5 mL 1% NaCl
    (3)   5 mL 1% $Al_2(SO_4)_3$
    (4)   5 mL 1% Na_2SO_4
    (5)   5 mL 1% $FeCl_3$

Stir each test tube thoroughly.   Record your observations at the time intervals indicated in the laboratory record.

NS   no settling, still muddy
BS   beginning to settle, still murky
SS   some settling, cloudy
MMS mostly settled, slightly cloudy

EXPERIMENT 23   WATER IN FOOD, ELECTROLYTES AND WATER PURITY

LABORATORY RECORD

PART I: PERCENT WATER IN A FOOD

A.   DATA

A-1   Type of food   _____

A-2   Mass of watch glass   _____

A-3   Mass of watch glass and food   _____

A-4   Mass of watch glass and dried food   _____

B.   CALCULATIONS

B-1   Mass of the food slices   _____

B-2   Mass of dried food   _____

B-3   Water lost by dehydration   _____

B-4   Percent water in the food item   _____
         Calculations:

B-4   Foods tested   % Water(Experimental)   Handbook value

         _____

         _____

         _____

         _____

         _____

         _____

PART II: ELECTROLYTES AND CONDUCTIVITY
C.   ELECTROLYTES AND CONDUCTIVITY

Substance	C-1 Type of Electrolyte (check one)			C-2 Particles in Solution (ions, molecules, both)
	strong	Weak	None	
distilled				
water				
mineral water				
NaCl(s)				
NaCl(aq)				
HCl(aq)				
NaOH(aq)				
HAc(aq)				
sugar(sucrose)				
glucose				
$NH_4OH(aq)$				
ethanol				

C-3   Write equations for the reaction of the following solutes   with water:

HCl(g)_____

NaOH(s)_____

HAc(l)_____

sugar, $C_{12}H_{22}O_{11}(s)$_____

glucose, $C_6H_{12}O_6$ (s)_____

180

D.   ELECTROLYTES IN BODY FLUIDS

Type of IV solution (D-1)

_____   _____   _____

Label (D-2)

Listing of electrolytes(meq/L)

cations(+)	anions(-)	cations(+)	anions(-)	cations(+)	anions(-)
_____	_____	_____	_____	_____	_____
_____	_____	_____	_____	_____	_____
_____	_____	_____	_____	_____	_____
_____	_____	_____	_____	_____	_____
_____	_____	_____	_____	_____	_____

Total charge

positive	negative	positive	negative	positive	negative
_____	_____	_____	_____	_____	_____

Overall charge

_____   _____   _____

QUESTIONS

State whether each of the following equations represents a weak
electrolyte, a strong electrolyte, or a nonelectrolyte?

1. HX $\underset{\longleftarrow}{\overset{H_2O}{\longrightarrow}}$ H^+(aq) + X^-(aq)

2. XY_2 $\overset{H_2O}{\longrightarrow}$ X^{2+}(aq) + $2Y^-$(aq)

3. XYZ(s) $\longrightarrow$ XYZ(aq)

4. YOH(s) $\longrightarrow$ Y^+(aq) + OH^-(aq)

EXPERIMENT 23   WATER IN FOOD, ELECTROLYTES AND WATER PURITY

PART III: TESTING WATER HARDNESS

(E-1)WATER SAMPLE   ML SOAP SOLUTION   WATER SAMPLE   ML SOAP SOLUTION

Distilled water   _____   Mineral water   _____

Tap water   _____   Sea water   _____

Hard water   _____   _____   _____

Soft water   _____   _____   _____

Which water sample was the softest water?   Why?

Which water sample was the hardest water?   Why?

F. PURIFICATION OF WATER BY SETTLING METHODS

TREATMENT CHEMICAL

TIME(MIN)	H_2O	NaCl	$Al_2(SO_4)_3$	Na_2SO_4	$FeCl_3$
0					
2					
5					
15					
30					
60					

How does the presence of ions affect the rate of settling compared to the rate of settling with water?

Which chemical produced the most rapid settling?

EXPERIMENT 24   CONCENTRATION OF A SOLUTION

PURPOSE

1. Evaporate a salt to dryness.
2. Calculate the percent(weight/volume) concentration of a salt solution.
3. Calculate the molar concentration of a salt solution.
4. Dilute a sample of the salt solution, evaporate to dryness, and calculate the percent(weight/volume) and molar concentration of the diluted sample.

MATERIALS

ringstand and iron ring
wire screen
two(2) evaporating dishes
NaCl or other salt solution (s)
400-mL beaker
small graduated cylinders, 10 mL, or 5-mL pipettes

KEYED OBJECTIVES IN TEXT: 9-1, 9-2, 9-3, 9-4, 9-5

DISCUSSION OF EXPERIMENT:

The percent (weight/volume)concentration of a solution states the amount of solute present in a specified volume of the solution.

$$\text{percent concentration} = \frac{\text{grams of solute}}{100 \text{ mL solution}}$$

A molar concentration describes the solute in moles and the volume of the solution in liters.

$$\text{molarity (M)} = \frac{\text{moles of solute}}{\text{liters of solution}}$$

In this experiment, you will measure out a known volume of a sodium chloride solution. The sample will be evaporated to dryness and the mass of the salt determined. Then you will be able to calculate the concentration of the initial solution.

An aqueous solution is diluted by adding water. This increase in volume means that the solute is less concentrated and the concentration of the solution (percent or molarity) will be lower than the original solution.

LABORATORY ACTIVITIES:

PART I: CONCENTRATION OF A SODIUM CHLORIDE SOLUTION

A.   DATA

A-1   Weigh two, dry evaporating dishes.   Record the mass of each.

A-2   Measure a 5.0-mL sample of NaCl solution using a 10-mL
      graduated cylinder. Record.   Pour the NaCl sample into the
      first evaporating dish, Measure out another 5.0 mL sample of
      NaCl solution and pour that sample into the second evaporating
      dish. Now add 5.0 mL of water to dilute this solution. You
      now have a 10.0 mL solution in the second evaporating dish.

A-3   WEAR YOUR GOGGLES FOR THE FOLLOWING PROCEDURE.   Fill the 400-
      mL beaker about one-half full of water and set on a wire
      screen. Set the evaporating dish on top of the beaker. Heat
      the water in the beaker to boiling. See Figure 24-1. You may
      need to replenish the hot water bath as you proceed. Repeat
      the same procedure with your diluted sample.

Figure 24-1.   Apparatus for evaporation of salt solution.

When the NaCl is almost dry, **carefully** remove the evaporating
dish, dry the bottom, and set the dish directly on the wire
screen. Heat gently with a low flame to completely dry the
salt.   Cool for 10 minutes. Weigh the evaporating dish and the
dried NaCl.   Repeat the same procedure with the diluted sample.

B.   CALCULATIONS

B-1   Calculate the mass of the NaCl after drying.

B-2   Calculate the percent(weight/volume) concentration.
       percent   = $\dfrac{\text{mass of NaCl(dried)}}{\text{volume of sample}}$ x 100
      concentration

B-3   Calculate the number of moles of NaCl (58.5 g/mol).

          moles NaCl   =   g NaCl   x $\dfrac{1 \text{ mol}}{58.5 \text{ g}}$

B-4   Convert the volume of the sample to liters.

B-5   Calculate the molarity of the NaCl solution.

          molarity = $\dfrac{\text{moles NaCl}}{\text{liters of solution}}$

184

EXPERIMENT   24   SOLUTION CONCENTRATION AND DILUTION

LABORATORY RECORD

A.   CONCENTRATION OF A SODIUM CHLORIDE SOLUTION

A.   DATA   ORIGINAL SOLUTION   DILUTED SOLUTION

A-1   Mass of evaporating dish   _____   _____

A-2   Volume of NaCl solution   _____   _____

A-3   Mass of dish and dried NaCl_____   _____

B.   CALCULATIONS

B-1   Mass of the dried NaCl   _____   _____
      Calculations:

D-2   Percent concentration   _____   _____
      Calculations:

D-3   Number of Moles of NaCl_____   _____
      Calculations:

D-4   Volume of sample in liters_____   _____

D-5   Molarity of the NaCl solution_____   _____
      Calculations:

EXPERIMENT 24   WATER IN FOOD AND FORMATION AND HEATS OF SOLUTIONS

QUESTIONS AND PROBLEMS

1. Describe the ways the body loses water and gains water.

2. 10.0 mL of a NaCl solution is placed in an evaporating dish and
   evaporated to dryness. The residue has a mass of 2.52 g.

   What is the percent(weight/volume) concentration of the solution?

   What is the molarity of the solution?

3. What is the percent(wt/vol) concentration of a solution prepared
    by dissolving 60 grams of KCl in water to make 400 mL of solution?

4. A 5% glucose solution is a typical solution used in the hospital.
    How many grams of glucose are in 500 ml of a 5% glucose solution?

5. What is the molarity of a solution that contains 80.0g of NaOH
   (molar mass = 40.0 g/mol) dissolved in 500 mL of solution?

EXPERIMENT 25   TESTING FOR IONS

PURPOSE

1. Identify a chemical reaction as an indication of the presence
    of a cation (positive ion) or anion (negative ion).

2. Use a chemical reaction to determine the presence of a
    cation or anion in samples such as juices, milk, bonemeal and
    an unknown salt.

MATERIALS

test tubes
test tube rack
small beakers
ringstand and iron ring
filter paper
flame-test wire
spot plate

solutions: (0.1M) $NaCl$, $AgNO_3$, $BaCl_2$, Na_3PO_4, Na_2SO_4,
              KCl, Na_2CO_3, $CaCl_2$, $FeCl_3$, $KSCN$, NH_4NO_3
              (6M or 6N) HCl, $NaOH$, HNO_3

KEYED OBJECTIVES IN TEXT: Chapter 3: Ions and ionic formulas
                                 Chapter 4: Chemical change, reactions

DISCUSSION OF EXPERIMENT

Many solutions such as the milk and the juices you drink contain an
assortment of ions due to the ionic compounds dissolved in those
solutions. By performing chemical reactions you will see observable
changes that are due to the presence of a particular ion.   In this
experiment, you will observe chemical changes such as the formation
of a solid(precipitate) a change in color, the formation of
bubbles(gas), or the color of the solution in a flame test.   Your
observations will be the key to identifying those same ions when
you test unknown solutions.   These are the kinds of procedures that
occur in the medical laboratory when identification of certain
components of urine or the blood must be made.   The following ions
will be tested:

Cations Na^+, K^+, Ca^{2+}, Fe^{3+}, NH_4^+

Anions Cl^-, PO_4^{3-}, SO_4^{2-}, CO_3^{2-}

PART I: TESTS FOR CATIONS Na^+, K^+, Ca^{2+}, NH_4^+, Fe^{3+}

In many parts of this experiment, you will be using about 2 mL portions of different solutions. Be sure to label each test tube and beaker and keep them in order. Remember that once you have removed a solution from the reagent bottle, it must be discarded even if you don't use all ot it. **NEVER RETURN CHEMICALS TO THEIR STORAGE BOTTLES.** You may wish to prepare small beakers of 6 M HCl, 6 M HNO_3, 6 M NaOH and 0.1 M $AgNO_3$. Another beaker of clean water is convenient for rinsing out the medicine dropper.

A. Sodium ion, Na^+

Place about 2 mL (40 drops) of 0.1 M NaCl in a test tube. Add 2-3 drops of 6 M HCl. (**Be sure you have placed a small amount of 6 M HCl in the beaker. Do not place a dropper in the 6M HCl reagent bottle.**) Clean a flame test wire by dipping it in a beaker containing a small amount of 6M(or 6N) HCl and placing the loop of the wire in a flame. Repeat the cleaning process until only a light blue color forms in the flame. Using a clean flame test wire, test the NaCl solution and record the color of the flame. The NaCl solution should give a bright, yellow-orange flame.

B. Potassium ion, K^+

A flame test is also used to identify the presence of K^+. However, the flame is not as persistent nor intense as Na^+. Place about 2 mL of 0.1 M KCl in a test tube. Add 2-3 drops of 6 M HCl. Clean the flame test wire, and then test the KCl solution. The color of a potassium flame is lavender and does not last long, so you must look for it immediately upon heating the flame test wire. Record the color you observe for the potassium ion, K^+.

C. Calcium ion, Ca^{2+}

Place about 2 mL(40 drops) of 0.1M $CaCl_2$ in a test tube. Using a small beaker, obtain a small amount of 0.1 M $(NH_4)_2C_2O_4$, ammonium oxalate. With a clean dropper, add 5-10 drops of ammonium oxalate to the test tube containing the $CaCl_2$. A slow formation of a cloudy, white solid(precipitate) is CaC_2O_4. This confirms the presence of Ca^{2+} in a sample.

$$Ca^{2+} \quad + \quad C_2O_4^{2-} \quad \longrightarrow CaC_2O_4(s)$$
$$\text{white}$$

Calcium also gives a flame test. Place about 2 mL 0.1 M $CaCl_2$ in a spot plate. Clean your flame test wire, and test the $CaCl_2$ solution for a red-orange color. Record.

D.   Ferric ion, Fe^{3+}

Place about 2 mL 0.1 M $FeCl_3$ in a clean test tube. Obtain a small amount of KSCN, potassium thiocyanate, in a beaker. Using a clean medicine dropper, add 2-3 drops of KSCN to the sample. A deep red color should appear. A faint pink color is not considered to be a positive test for iron. Record results.

$$Fe^{3+} + SCN^- \longrightarrow FeSCN^{2+}$$
$$\text{deep red}$$

E.   Ammonium ion, NH_4^+

Place about 2 mL of 0.1 M NH_4NO_3 in a test tube. Add 2-3 drops 6 M NaOH, sodium hydroxide, to the test tube. Place a strip of moistened red litmus paper in the top of the test tube. Attach a test tube holder and gently heat the solution. **DO NOT BOIL.**   As $NH_3(g)$, ammonia gas, is formed, it rises and turns the litmus paper blue.   The odor of ammonia may be noticed if you carefully waft the vapors from the test tube towards you. **DO NOT PLACE THE TEST TUBE DIRECTLY UNDER YOUR NOSE.**

$$NH_4^+ + OH^- \longrightarrow NH_3(g) + H_2O$$
$$\text{ammonia}$$

PART II: Cl^-, PO_4^{3-}, SO_4^{2-}, CO_3^{2-},

F.   Chloride ion, Cl^-

Place about 2 mL of 0.1 M NaCl in a test tube. Add 3-4 drops of 0.1 M $AgNO_3$. **BE CAREFUL. $AgNO_3$ STAINS THE SKIN.** Using a medicine dropper, add 3-4 drops of $AgNO_3$ to the solution in the test tube.   You should see the formation of a white precipitate, AgCl.

$$Ag^+ + Cl^- \longrightarrow AgCl(s)$$
$$\text{white}$$

You can confirm the presence of AgCl by adding 10-15 drops of 6 M HNO_3 to the test tube with the white precipitate.   The AgCl precipitate <u>will not</u> dissolve.

G. Phosphate ion, PO_4^{3-}

Place about 2 mL of 0.1 Na_3PO_4 in a test tube. Add 3-4 drops of 0.1 M $AgNO_3$. CAUTION. A yellow precipitate of Ag_3PO_4 should form.

$$3\ Ag^+\ +\ PO_4^{3-}\ \longrightarrow\ Ag_3PO_4(s)$$
$$\text{yellow}$$

You can confirm the presence of Ag_3PO_4 by adding 10-15 drops of 6 M HNO_3 to the test tube. Stir with a glass rod. The yellow precipitate will dissolve.   You may need to add a few more drops of 6 M HNO_3 to completely dissolve the Ag_3PO_4.

H. Sulfate ion, SO_4^{2-}

Place about 2 mL of 0.1 M Na_2SO_4 in a test tube. Add about 10 drops of 6 M HCl. Obtain a small amount of 0.1 M $BaCl_2$. Add 10-15 drops of $BaCl_2$ to the solution in the test tube. A fine, white precipitate of $BaSO_4$ should appear. Record results.

$$Ba^{2+}\ +\ SO_4^{2-}\ \longrightarrow\ BaSO_4(s)$$
$$\text{white}$$

I. Carbonate ion, CO_3^{2-}

Place about 2-3 mL of Na_2CO_3 in a test tube. While carefully observing the solution, add a few drops of 6 M HCl to the test tube. Watch for a strong evolution of gas(bubbles) as you add the HCl.

$$CO_3^{2-}\ +\ 2\ H^+\ \longrightarrow\ CO_2(g)\ +\ H_2O$$
$$\text{bubbles}$$

PART III: IDENTIFICATION OF CATIONS AND ANIONS IN UNKNOWN SALTS AND SOLUTIONS

J.   IDENTIFICATION OF AN UNKNOWN SALT AND ITS FORMULA

J-1   A salt is composed of one cation and one anion. By carrying out the same tests for cations and anions, you can identify the particular cation and anion that make up your unknown salt. Obtain your sample of an unknown salt from your instructor. Record the sample number.

J-2   If the unknown sample is a solid, dissolve 1 g of the salt in 50 mL of water. This is now your sample solution ready for testing with the same reagents you used in previous tests.

Begin your testing with the cations by doing some flame tests. Record the color, if any, observed in the flame test. To test for Ca^{2+}, use 2 mL of your sample solution you prepared, and add some $Na_2C_2O_4$. Compare the results with the previous test you did for Ca^{2+}. Repeat the tests for Fe^{3+} and NH_4^+ by using about 2 mL of your sample solution prepared from your unknown salt. Record results.

The anions will be testing in a similar way. Begin by placing 2 mL of your sample solution of the unknown salt in a test tube. To test for Cl^-, add some $AgNO_3$ as you did before. Record your observations. Continue with additional 2 mL samples of the sample solution to test for the presence of other anions.

J-3   When you check your results of testing the unknown salt solution, there should be an indication of the presence of one cation (sodium, potassium, calcium, iron or ammonium) and one anion(chloride, phosphate, sulfate, or carbonate. For example, if your unknown salt was $CaCl_2$, you would have noticed a positive test for Ca^{2+} and a positive test for Cl^- as you were testing the sample solution. Write the ions you identified through the testing.

J-4   Use ionic charges to write a formula and name of your salt.

K.   **TESTING SOLUTIONS FOR THE PRESENCE OF IONS**

The same tests may also be applied toward the identification of ions in sample of juices and liquids. Your instructor will tell you what samples from the following list you are to test. Use 2 mL of the sample to test for each of the ions. Report your results.

Juices   Obtain 25-30 mL of a fruit juice. If it contains pulp, filter through cheese cloth first.

Sodas   Obtain 25-30 mL of a soft drink or mineral water. For colas, root beers, or others with deep colors, mix with a small amount of charcoal in a small beaker. Charcoal will absorb the pigments. Filter and use the filtrate for testing.

Bone   A scoop of bone meal, or small chicken or fish bones, or teeth, can be dissolved in 15 mL distilled water and 10 mL 6 M HNO_3 in a beaker. Heat gently until the material dissolves.   Allow to cool. Filter.

Milk   Obtain 30 mL of nonfat milk. Add 10 mL of 0.1M HAc, acetic acid. Small curds of protein will form. ilter using cheesecloth. Reduce the volume to 25 mL by heating. Cool and filter. Use 2 mL to test.

PART I: TESTING FOR CATIONS

Ion	Test	Observations	Product
A. Na^+	flame test	Na^+ produced a bright yellow flame.	
B. K^+			
C. Ca^{2+}			CaC_2O_4
D. Fe^{3+}			
E. NH_4^+			

PART II: TESTING FOR ANIONS

F. Cl^-			
G. PO_4^{3-}			
H. SO_4^{2-}			
I. CO_3^{2-}			

PART III: IDENTIFICATION OF AN UNKNOWN SALT AND ITS FORMULA

J-1   Unknown Sample Number _____

J-2   Indicate the tests that were performed with the unknown sample
      solution. Give your observations and conclusions about the
      presence or absence of the ion you tested.

Test	Procedure	Observation	Ion Present, if any

J-3 **SUMMARY OF TEST RESULTS**

Cation present in the unknown salt   _____

Anion present in the unknown salt   _____

J-4

Formula of the salt   _____

Name of the salt   _____

K.   TESTING OF SOLUTIONS

Type of solution tested _____

Test Procedure	Observation	Ion present, if any

SUMMARY OF SOLUTION TESTING:

Cations present in the solution: _____

_____

Anions present in the solution: _____

_____

EXPERIMENT 26   FLUID TRANSPORT

PURPOSE

1. Observe the transport systems in solutions.
2. Determine the ability of suspension particles, colloids, and
    true solution particles to pass through filters and/or membranes.
3. Distinguish between osmosis and dialysis.

MATERIALS

A. beaker
    test tubes
    test tube rack
    10% starch solution
    10% NaCl
    iodine reagent, charcoal
    funnel
    filter paper(Whatman #1)
    0.1M AgNO$_3$
    Benedict's reagent
    1 M glucose

B. 20-cm dialysis bag
    5% starch
    charcoal
    250-mL beaker
    50-mL graduated cylinder
    iodine reagent
    funnel, filter paper

C. 20-cm dialysis bag
    distilled water
    solutions: 1M glucose
       0.1 M AgNO$_3$
       10% NaCl
       Benedict's reagent
       10% starch
    beakers, 100 and 250-mL
    test tubes (8)
    funnel
    boiling water bath
    iodine reagent

KEYED OBJECTIVES IN TEXT: 9-6, 9-7, 9-8, 9-9

DISCUSSION OF EXPERIMENT

Materials move in and out of cells by (1) **filtration** produced by
gravity; (2) **osmosis** whereby water moves across a semipermeable
membrane; and (3) **dialysis** whereby true-solution particles and
water move across a semipermeable membrane when concentration
gradients exist.

     In Experiment B, you will see the separation of suspension
particles from colloids by filter paper.

     In Experiment C, osmosis will be demonstrated by the movement
of water through a semipermeable membrane while colloidal particles
of starch are retained within the membrane. Water diffuses from
its area of greater concentration (outside) to an area of lower
concentration (inside). This means that water flows in the
direction of greater solute concentration.

In Experiment D, dialysis will be demonstrated by the movement of true-solution particles through a semipermeable membrane while coloidals are retained. Many of the membranes in the body are dialyzing membranes. For example, the intestinal tract uses a semipermeable membrane that allows the simple, true-solution-sized particles from digestion to pass into the blood and lymph. Larger, incompletely digested food particles(colloids) are retained with the intestinal tract. Dialyzing membranes are also used in hemodialysis to separate waste particles, particularly urea, out of the blood.

LABORATORY ACTIVITIES

A. IDENTIFICATION TESTS

Throughout this experiment, you will need to test for the presence of certain substance through identification tests. After you complete this section, you can refer to its results for determining whether a particular substance is present or absent in a solution.

A-1 In a small beaker, prepare a mixture of 15 mL of 1M glucose, 15 mL of 10% NaCl, and 15 mL of a 10% starch solution. After stirring the mixture, prepare three test tubes by pouring 5 mL of the mixture into each test tube. SAVE THE REMAINING MIXTURE for the dialysis experiment later. Carry out the following tests and record results.

Test tube 1   Test the mixture for Cl⁻ by adding 5 drops of 0.1M $AgNO_3$.

Test tube 2   Test the mixture for starch by adding 3-4 drops of iodine reagent.

Test tube 2   Test the mixture for glucose by adding 5 mL of Benedict's reagent and heating the test tube and contents in a boiling water bath for 5 min.

B. FILTRATION

B-1 In a small beaker, prepare a mixture of 5 mL of the 5% starch solution and a small amount of charcoal, and 10 mL water. Pour the mixture into a funnel fitted with a filter paper. Collect the liquid(filtrate) that passes through the paper into the test tube. Observe the filter paper and the filtrate for evidence of the charcoal. Record your observations.

B-2   Test the filtrate for starch by adding a few drops of iodine reagent to the liquid in the test tube.   Record your observations.

B-3   Identify the particles of starch and charcoal as colloids or suspensions.

C. DIALYSIS

C-1   Obtain a dialysis bag that has been softened in distilled water, tie a knot in one end.   Place a funnel in the open end and pour in the remaining mixture you prepared in step A-1.   Tie a firm knot in the other end to close the dialysis bag. Rinse the dialysis bag with distilled water.   Place the dialysis bag in a 250-mL beaker containing 100-mL distilled water.   See Figure 26-1.

Immediately pour off 15 mL of the water surrounding the bag for the first group of tests.   Divide this sample into three test tubes (5 mL each).   Test the first(1) solution for Cl^-, the second (2) for starch and the third for glucose. (Use the identification tests.)   Record your results for time 0 min as present(+) or absent(-).

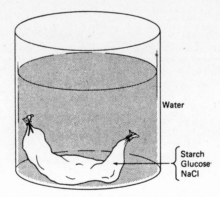

Figure 26-1   Dialysis bag placed in distilled water.

C-2   In 30 min, remove another 15-mL sample of the water in the beaker, separate into three test tubes, and repeat the same three tests again.   Record results.

C-3   Repeat the tests at 60 min and record results.   You may wish to leave the dialysis bag in the distilled until the next laboratory rechecking all the tests again.

C-4   Break the bag open and test the remaining contents for all three substances.   Record results.

197

EXPERIMENT 26   FLUID TRANSPORT   NAME_____
LABORATORY RECORD   DATE_____
                                               SECTION_____

A.   IDENTIFICATION TESTS

                                   Test Result

          Cl$^-$ _____

          starch   _____

          glucose   _____

B.   FILTRATION

B-1   Appearance of filter paper   _____

      Substance present   _____

B-2   Appearance of filtrate   _____

      Appearance of filtrate after
         iodine reagent   _____

      Substance present   _____

Indicate the type of particles represented by charcoal and starch

          Particle   Suspension   Colloid

          charcoal   _____   _____

          starch   _____   _____

C. DIALYSIS

C-1 Record your observations after testing. Decide if any substance was present in the water surrounding the dialysis bag. Indicate the presence(+) or absence(-) of each substance.

Time Sample Removed	Cl⁻	starch	glucose
0 min			
30 min			
60 min			
Next lab			
Contents remaining inside dialysis bag			

Answer the following questions by referring to the last test results (60 minutes or the next lab.)

1. Which substances were found in the water surrounding the dialysis bag?

2. Why were those substances able to get through the dialysis bag?

3. What substances remained inside the dialysis bag ? Why?

QUESTIONS AND PROBLEMS

1. Osmosis occurs through the walls of red blood cells. When a red blood cell is placed in a solution whose solute concentration is **isotonic**, the same as that inside the red blood cell, the flow of water is equal into and out of the cell. The effect of osmosis is seen when a cell is placed in a solution with a lower (**hypotonic**) or a higher (**hypertonic**) solute concentration. Both 0.9% NaCl (saline) and 5% glucose solutions are considered isotonic to the cells of the body. State whether the following concentrations of various solutions are isotonic, hypotonic, or hypertonic:

a. H_2O _____

b. 0.9% NaCl   _____

c. 10% glucose   _____

d. 3% NaCl   _____

e. 0.2% NaCl   _____

2. A red blood cell in a hypertonic solution will shrink in volume(undergo creation); a red blood cell in a hypotonic solution will swell and possible burst (undergo hemolysis). Predict the effect on the volume when a red blood cell is placed in the following solutions:

a.   2% NaCl   _____

b. H_2O _____

c.   5% glucose   _____

d.   1% glucose   _____

e.   10% glucose   _____

3.   Why are only isotonic solutions used as parenteral solutions? (A parenteral solution is any solution not given orally.)

EXPERIMENT 27   ACIDS, BASES AND pH

PURPOSE

1. Use a naturally occurring dye as a pH indicator.
2. Use a pH meter and an indicator to determine the pH of several substances.
3. Observe the changes in pH as acid or base is added to buffered and nonbuffered solutions.

BRING FROM HOME: Samples to test for pH (shampoo, conditioner, mouth wash, etc.)

MATERIALS

A. red cabbage
    400-mL beaker
    distilled water

B. pH meter
    buffer set(pH 1-13)
    Solutions:
       0.1M HCl, 0.1M NaOH

D. buffers set

C. pH meter
    test tubes
    Samples to Test: vinegar,
    lemon juice, ammonia, juices,
    shampoos, conditioners, etc.
    shell vials

D. Set of buffers
    0.1 M NaCl

KEYED OBJECTIVES IN TEXT: 10-1, 10-2, 10-3, 10-4,10-5,10-6, 10-7

DISCUSSION OF EXPERIMENT:

Certain natural substances contain pigments that change color when the pH changes. The pigment found in red cabbage leaves is such a substance. By extracting(removing) this pigment from the leaves of the red cabbage to form an aqueous solution, we can prepare a natural indicator for pH determinations.

You will first set up a series of reference solutions of known pH values from 0 to 14. When you add the cabbage pigment solution, different colors will develop. Note the color and pH of the buffer solution you placed in that test tube. This will give you a reference set of pH colors, one for each pH value from 0 to 14.

The pH of a solution is a measure of the acidity, neutrality, or basicity of that solution. A pH value in the range of 0 to 7 is acidic; a pH of 7 is neutral; and a pH greater than 7 (7 to 14) is basic.

pH scale

0 1 2 3 4 5 6 7 8 9 10 1 12 13 14

very acidic   neutral   very basic

EXPERIMENT 27   ACIDS, BASES AND pH

The pH of the blood is maintained between 7.35 and 7.45 by buffers in the body. Buffers utilize a system of a weak acid or base and its salt. The effects of compounds in the body that would change the pH of the blood and body fluids are offset by the buffer system.

LABORATORY ACTIVITIES

A. PREPARATION OF A NATURAL INDICATOR

Tear up a few leaves of a red cabbage and place them in the bottom of the 400-mL beaker. Add 100 mL of distilled water to cover the leaves. Place the beaker on a wire gauze on the ring stand set up. Heat the water, but do not let it boil vigorously. In 5-10 min, the solution should become a deep purple color. If not, add another 1-2 leaves of red cabbage and heat again. Cool. While the dye is being extracted, prepare your set of reference pH standards in Part B. The cabbage indicator will be used in experiments B, C, and D.

B. PREPARATION OF A pH REFERENCE STANDARD

To set up the pH reference standard, you need 14 test tubes. This may be accomplished by combining your test tube set with a partner's set. Place 3-4 mL of 1 M HCl in the first test tube. This will be pH 0. For pH values of 1 to 13, use 3-4 mL of the proper buffer from the pH buffers available. In the final test tube, place 3-4 mL of 1 M NaOH. This will be pH 14. You should now have test tubes arranged in order of pH 0 through 14.

When the cabbage solution is prepared and cooled, add 2-3 mL(40-60 drops) of the cabbage indiciator to each of the solutions in the pH reference set. If you wish a deeper color, use another 1 mL of the cabbage indicator. Record the colors of the pH solutions. **Keep** this reference set for the next experiment.

C. DETERMINATION OF pH

Using Cabbage Indicator Obtain 3-4 mL of one of the solutions whose pH you are to determine. Add about 2 mL(40 drops) of cabbage indicator to the solution. Compare the resulting color of the unknown solution to the colors in your pH reference set. Find the reference solution that has the closest match in color to the unknown. The pH of the reference solution will also be the pH of the unknown. Test for the pH of several solutions available in the lab.   Record.

pH Meter If a pH meter is available, it may be used to determine the pH values of the unknown solutions. In addition samples with strong colors that cannot be used with cabbage indicator, such as coffee and cola, may be checked with the pH meter.   Test solutions for pH and record.

EXPERIMENT 27   ACIDS, BASES AND pH

D.   EFFECT OF BUFFERS UPON pH

D-1   Place 5 mL of each of the following solutions into four
      shell vials (or test tubes):

      1. H_2O

      2.   0.1M NaCl

      3.   A buffer with a high pH

      4.   A buffer with a low pH

      Determine the initial pH of each solution. (Use a pH
      meter, or add 2 mL of cabbage indicator.) Record.

D-2   Add 5 drops of 0.1M HCl (acid) to each of the solutions.
      Stir and determine the pH again of each solution.

D-3   Determine the change in pH units, if any. Record.

D-4   Set up four shell vials (or test tubes) as you did in step
      D-1 using the same four solutions. Add 2 mL of cabbage
      indicator to each, and report the intial pH of each solution.

D-5   Add 5 drops of 0.1M NaOH(base) to each of the solutions.
      Stir and record the pH of the solutions.

D-6   Determine the change in pH. Record.   Answer the questions
      in the laboratory record sheet.

EXPERIMENT 28   ACIDS, BASES AND BUFFERS   NAME _____

LABORATORY RECORD

SECTION_____

DATE_____

A., B.   <u>pH</u> <u>REFERENCE</u> <u>COLOR</u> <u>OF</u> <u>CABBAGE</u> <u>INDICATOR</u>

pH	Color	pH	Color
0	_____	8	_____
1	_____	9	_____
2	_____	10	_____
3	_____	11	_____
4	_____	12	_____
5	_____	13	_____
6	_____	14	_____
7	_____		

Complete the table:

$[H^+]$	$[OH^-]$	pH	acid, base, neutral
1×10^{-6}			
		10	
			neutral
	1×10^{-3}		
		2	

207

C. DETERMINATION OF pH

Solution	Color	pH	pH Using pH meter	Acidic, basic, or neutral?
cleaners				
vinegar	_____	___	_____	_____
ammonia	_____	___	_____	_____
_____	_____	___	_____	_____
juices				
7-UP	_____	___	_____	_____
_____	_____	___	_____	_____
_____	_____	___	_____	_____
detergents				
_____	_____	___	_____	_____
_____	_____	___	_____	_____
shampoos				
_____	_____	___	_____	_____
_____	_____	___	_____	_____

Other samples tested

_____	_____	___	_____	_____
_____	_____	___	_____	_____
_____	_____	___	_____	_____
_____	_____	___	_____	_____

D. EFFECT OF BUFFERS UPON pH

	H_2O	NaCl	Buffer (high pH)	Buffer (low pH)
Reactions with HCl				
D-1 pH(initial)				
D-2 pH(final)				
D-3 pH change				
Reactions with NaOH				
D-4 pH(initial)				
D-5 pH (final)				
D-6 pH change				

QUESTIONS

Which solution(s) showed the greatest change of pH? Why?

Which solutions(s) showed little or no change in pH?   Why?

What is the function of a buffer?

What components are necessary to make up a buffer system?

QUESTIONS AND PROBLEMS

1. The label on the shampoo claims that it is pH balanced. What
    do you think "pH balanced" means?

2. Why would you expect vinegar to be more acidic than milk?

3. A sample of 0.020 mole HCl is dissolved in water to make a
    2000 mL solution. Calculate the molarity of the HCl solution;
    the [H$^+$], and the pH.

4. Normally, the pH of the human body is fixed in a very narrow
    range between 7.35 and 7.45   A patient with a acidotic
    blood pH of 7.3 may be treated with alkali such as sodium
    bicarbonate. Why would this treatment raise the pH of the
    blood?

5. What kinds of changes would you expect in pH when a small
    amount of base is added to a buffered solution and an
    unbuffered solution?

EXPERIMENT 28   ACID-BASE TITRATIONS

PURPOSE

1. Prepare a sample for titration with an acid or a base.
2. Correctly reach an endpoint for an indicator.
3. Calculate the percentage of acid in a product.
4. Compare the acid-absorbing capacity of several antacids.

MATERIALS

I. vinegar (white)
    250-mL Erlenmeyer flask
    10-ml graduated cylinder
    phenolphthalein indicator
    0.1 M NaOH
    50-mL buret

II. 250-mL Erlenmeyer flask
    antacid samples
    0.1 M HCl
    0.1 M NaOH
    phenolphthalein indicator
    50-mL buret

KEYED OBJECTIVES IN TEXT: 10-7

DISCUSSION OF EXPERIMENT

Vinegar is an aqueous solution of acetic acid (HAc). You will determine the percentage of acetic acid in a vinegar sample by titrating the acid with a known concentration of sodium hydroxide, NaOH. In the neutralization reaction, the acidic hydrogen of the acetic acid is neutralized by the hydroxide ion of the base to form water.

$$CH_3COOH \ + \ NaOH \ \longrightarrow \ CH_3COO^-Na^+ \ + \ H_2O$$
acetic acid

As the pH of the solution changes in the process of neutralization, the color of an indicator also changes. When neutraliation has been achieved with phenolphthalein, the solution will turn from a clear color to a faint, but permanent pink color. When you reach this point, stop adding base. This is called the end point of the titration.

By measuring the volume of the NaOH added and knowing its molarity from the label on the bottle, you can calculate the moles of NaOH that were used. At neutralization, the moles of NaOH equal the moles of acetic acid (HAc) present in the sample. Converting to the number of grams of HAc in the vinegar allows the calculation of the percentage of HAc in the vinegar.

   Stomach acid is primarily hydrochloric acid (HCl), about 0.1
M. Sometimes, when a person is under stress, excess HCl may be
produced, causing discomfort. An agent called an **antacid** is used
to absorb some of the acid. One commerical claims that its antacid
product consumes 47 times its own weight in excess stomach acid; 1
g of the antacid neutralizes 47 g of HCl in the stomach. You can
test such a claim by weighing an antacid tablet and reacting that
tablet with a solution of 0.1 M HCl. If you know how much HCl you
started with and how much HCl remains unreacted, you can calculate
the amount of HCl that was absorbed by the antacid. Several
antacid products may be tested.

LABORATORY ACTIVITIES

PART I: PERCENT CONCENTRATION OF ACETIC ACID IN VINEGAR

A.   TITRATION DATA

A-1 Using a 10-ml graduated cylinder or a 5.0 mL pipette, place
     5.0 mL vinegar in a 250-mL Erlenmeyer flask.   Record this
     volume.   Record the brand of vinegar.

     Add 25 mL of water to increase the volume of the solution
     for titration. This will not affect your results. Add 2-3
     drops of the phenolphthalein indicator to the vinegar solution
     in the flask.

A-2 Set up a buret as demonstrated by your instructor. Using a
     250 mL beaker, obtain about 100-150 mL of NaOH solution.
     Record the **molarity** of the NaOH solution as stated on the
     bottle. See Figure 28-1.

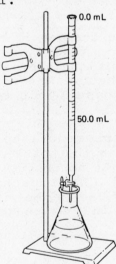
0.0 mL

50.0 mL

      Figure 28-1   Apparatus for acid-base titration.

Rinse the buret with 10 mL of NaOH. To fill the buret, set a funnel in the top, and slowly pour the NaOH from the beaker into the buret until the level of the NaOH is slightly above the 0.0 mL mark. (Be sure to lift up the funnel as you near the top.) Run the volume of NaOH down to the 0.0 mL mark by draining the NaOH in the buret into an extra beaker. The tip should now be full of NaOH solution and free of bubbles.

A-3   Record the initial level of the NaOH in the buret. This is usually 0.0 mL. Begin adding NaOH to the vinegar solution in the flask. (Be sure you have added indicator.) As you add the NaOH, swirl the solution in the flask to completely mix the acid and the base.   At first, the red streaks produced by the reaction will disappear quickly.   As you approach the end point of the neutralization, the red or pink streaks will disappear more slowly.   Slow down the addition of the NaOH to drops at this time.   SOON, ONE DROP OF NaOH WILL CAUSE THE ENTIRE SOLUTION TO TURN A FAINT, PERMANENT PINK COLOR.   STOP ADDING NaOH.   You have reached the end point of the titration. See Figure 28-2.

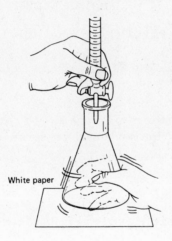

White paper

Figure 28-2 The solution in the flask is swirled as the sample is titrated.

A-4   The acid in the vinegar has been neutralized.   Record the level of the NaOH in the buret. Repeat the procedure two or three times using the same type of vinegar.

B. CALCULATIONS

B-1 Calculate the volume(s) of NaOH used to neutralize the vinegar sample.

B-2 Calculate the average volume of NaOH used.

$$\frac{volume(1) \ + \ volume(2) \ + \ volume(3)}{3} \ = \ average \ volume \ NaOH$$

B-3 Calculate the moles of NaOH using the volume and molarity of the NaOH.

$$moles \ NaOH \ = \ volume(L) \ NaOH \ x \ \frac{moles}{liter}$$

B-4 Record the moles of acid present in the vinegar (equal to the number of moles of NaOH).

$$moles \ HAc \ = \ moles \ NaOH$$

B-5 Calculate the molarity(M) of the acid.

$$\frac{moles \ HAc}{volume(L)} \ = \ M$$

B-6 Calculate the number of grams of acetic acid (HAc) in the vinegar sample. The molecular mass of HAc (CH_3COOH) is 60.0 g/mol.

$$g \ HAc \ = \ mol \ HAc \ x \ \frac{60.0 \ g}{mol}$$

B-7 The percentage (weight/volume) HAc can now be calculated.

$$\% \ HAc \ in \ vinegar \ = \ \frac{g \ HAc}{5.0 \ mL \ vinegar \ sample} \ x \ 100 \ \%$$

B-8 Observe the %HAc (acetic acid) listed on the label of the vinegar bottle. Record.   Calculate your percentage error.

$$\frac{difference \ between \ actual \ and \ experimental}{actual(label)} \ x \ 100 \ = \ \% \ error$$

OPTIONAL EXPERIMENT: Design a method to determine the amounts of acid in other substances such as soft drinks or fruit juices. In fruit juices, the acid is citric acid, M.W. 192.0 g/mol. Since citric acid has three acidic groups, its neutralization reaction requires 3 NaOH.

citric acid   +   3 NaOH ⟶ Na₃citrate + 3 H₂O

$$\text{citric acid} + 3\,NaOH \longrightarrow Na_3citrate + 3\,H_2O$$

$$
\begin{array}{l}
\quad\quad O \\
\quad\quad \| \\
CH_2-COH \\
\ |\ \ O \\
\ |\ \ \| \\
HO-C-COH \\
\ |\ \ \ O \\
\ |\ \ \ \| \\
CH_2-COH
\end{array}
\quad + \quad 3\,NaOH \longrightarrow
\begin{array}{l}
\quad\quad O \\
\quad\quad \| \\
CH_2-CO^- \\
\ |\ \ o \\
\ |\ \ \| \\
HO-C-CO^- \\
\ |\ \ \ O \\
\ |\ \ \ \| \\
CH_2-CO^-
\end{array}
\quad + 3Na^+ \quad + \quad 3H_2O
$$

You can choose juices such as lemon or grapefruit using a 25 mL to 50 mL sample. Record the volume used. The indicator phenolthalein(3-4 drops) can be used although the color of the fruit juice must be considered. For example, a yellow or orange colored juice will change to a red or reddish-orange color for the end point. Run at least two titrations with one of the fruit juices and calculate an average molarity and percent concentration.

PART II:TITRATION OF AN ANTACID

C.   TITRATION DATA

C-1   Record the active alkali ingredient listed on the label of the antacid package.

C-2   Carefully weigh a 250-mL Erlenmeyer flask. Record its mass. Crush one antacid tablet and transfer the crushed tablet to the Erlenmeyer flask. Weigh the flask and antacid material. Record.

C-3   Carefully measure out 100 mL of 0.1 M HCl, and add to the antacid in the flask. Allow the crushed material to dissolve. Heat gently over a moderate flame for 2-5 minutes. Some cloudiness will remain due to undissolved starches and binders.   Add 2-4 drops of phenolphthalein indicator. The solution should be colorless. If it turns pink, add another 25 mL of the 0.1M HCl. Record the total volume of 0.1 M HCl added to the antacid sample.

The antacid will neutralize a certain amount of the HCl. To determine this amount, we need to find out how much HCl remains that did not react with the antacid.

C-4   Fill a buret with NaOH to a level slightly above the 0.0 mL reading. Let the excess run out into a separate beaker to bring the volume of NaOH down to the 0.0 mL mark. Record the initial level of the NaOH solution.

C-5   Titrate the antacid solution by adding NaOH to the flask containing the dissolved antacid tablet and indicator. Stop titrating when the solution turns a permanent, faint pink color when you add one drop of NaOH. Record the volume level of the NaOH in the buret.

D.   CALCULATIONS

D-1   Determine the mass of the antacid substance. This will be the difference in the mass of the empty flask and the flask containing the crushed antacid tablet.

D-2   Calculate the volume of NaOH used in the titration. This will be equal to the volume of the HCl that did not react with the antacid. Record this value also as the volume of excess HCl.

D-3   Calculate the volume of HCl that reacted with the antacid.

HCl reacted =   100 mL HCl added   - volume(mL) of excess HCl

D-4   If we assume that 1 mL of the 0.1 M HCl has the same density as water (1.0 g/mL), we can convert the mL of HCl solution to grams of HCl that reacted with the antacid.

g HCl reacted   =   mL HCl reacted

D-5   The mass of HCl consumed by 1 g of the antacid can now be determined.

$$\frac{g\ HCl\ reacted}{g\ antacid} \quad = \quad g\ acid\ absorbed/g\ antacid$$

D-6   Record values obtained in the class for other antacids.

EXPERIMENT 28   ACID-BASE TITRATIONS   NAME_____
                                                SECTION_____
LABORATORY RECORD   DATE_____
EXPERIMENT 27   ACID-BASE TITRATIONS

PART I: DETERMINING THE PERCENT CONCENTRATION OF ACETIC ACID IN VINEGAR

A-1 Brand of vinegar _____
                          1st titration   2nd titration   3rd titration

       Volume   _____   _____   _____

A-2 Molarity of
     NaOH   _____   _____   _____

A-4 Final NaOH level_____   _____   _____

A-3 Initial NaOH level_____   _____   _____

CALCULATIONS

B-1 Volume of
     NaOH used   _____   _____   _____

B-2 Average volume NaOH
     Calculations: _____

B-3 Moles NaOH
     Calculations: _____

B-4 Moles HAc   _____

B-5 Volume of vinegar
     sample in liters   _____

     Molarity(M)   _____

B-6 g HAc   _____

B-7 % HAc in vinegar   _____
      Calculations:

B-8 % Error   _____
      Calculations:

OPTIONAL EXPERIMENT:

Design your own data page for determining the percent acid in your
choice of fruit juices. Show calculations where necessary.

PART II: DETERMINATION OF THE AMOUNT OF ACID NEUTRALIZED BY AN ANTACID

C.   TITRATION DATA   SAMPLE 1   SAMPLE 2

C-1   Brand of antacid   _____   _____

      Basic substance(s)   _____   _____
      listed on label

C-2   Mass of flask   _____   _____

      Mass of flask and antacid _____   _____

C-3   Volume of 0.10 M HCl added_____   _____

C-5   Final level of NaOH   _____   _____

C-4   Initial level of NaOH   _____   _____

CALCULATIONS

D-1   Mass of the antacid   _____   _____

D-2   Volume of NaOH used to _____   _____
      titrate excess HCl

      Volume of excess HCl   _____   _____

D-3   Volume of reacting HCl _____   _____

D-4   Grams of HCl solution
      reacting with antacid   _____   _____

D-5   Mass of HCl solution absorbed
      by 1 gram of antacid   _____   _____
      Calculations:

Summary of Absorbing Power of Various Antacids

D-6   Brand Antacid   Mass HCl/ gram antacid

      _____   _____

      _____   _____

      _____   _____

      _____   _____

EXPERIMENT 29   SOME PROPERTIES OF ORGANIC AND INORGANIC COMPOUNDS

PURPOSE

1. To observe differences in the solubility, flammability, melting and boiling points of organic and inorganic compounds.

2. To summarize physical characteristic most typical of organic and inorganic compounds.

polar - shows difference in charges.

MATERIALS

evaporating dish
test tubes
chemicals: pentane, NaCl(s), cyclohexane
chemistry handbook(s)

KEYED OBJECTIVE IN TEXT: 11-1

DISCUSSION OF EXPERIMENT:

The organic compounds share certain properties that can be distinguished from inorganic compounds. Many organic compounds contain covalent bonds between carbon and hydrogen with oxygen and/or nitrogen sometimes present. The general nonpolar nature of organic compounds makes them more soluble in nonpolar solvents than in water. Their melting and boiling points tend to be much lower than those of polar and ionic compounds.

LABORATORY ACTIVITIES

NOTE: ORGANIC COMPOUNDS ARE EXTREMELY FLAMMABLE. USE OF THE BUNSEN BURNER WILL BE LIMITED. DISPOSE OF ORGANIC COMPOUNDS IN SPECIAL CONTAINERS OR IN THE DRAINS FOUND IN THE HOODS.

A. FLAMMABILITY

Organic compounds are extremely flammable. Your instructor may do this experiment as a demonstration. Proceed with caution. There should be no Bunsen burners operating in the area.

   WORK IN THE HOOD. Test the compounds, pentane and NaCl, for flammability by placing very small amounts of each in an evaporating dish. Use 2-5 drops (NO MORE) of one of the pentane. Use a lighted match to burn the compound. Describe your observations. Report if the compounds are flammable (F) or nonflammable (NF).

B. SOLUBILITY

Test the solubility of the compounds, pentane and NaCl, in water and in cyclohexane, a nonpolar solvent. Place 2-3 mL of water, a polar solvent, in two test tubes. Add a small amount, 5 drops of the pentane, or a small amount of NaCl solid held on the tip of a

221

spatula, to the solvent. Stir the contents of each test tube.
Record your observations. Record whether the compounds are soluble
(S) or insoluble (I).

Repeat using 2-3 mL of cyclohexane as the solvent. Record your
observations.

C. MELTING AND BOILING POINTS

Obtain a chemistry handbook and look up the melting and boiling
points of NaCl and pentane.

D. SUMMARY

Complete the table summarizing the properties and characteristics
of organic and inorganic compounds using pentane as a typical
organic compound and NaCl as a typical inorganic compound.

nothing but single bonds (hydrocarbons - H and C)

PURPOSE

1. Construct models of some representative hydrocarbons.
2. Write the structural and condensed formulas of organic compounds whose models are on display.
3. Write the names of organic compounds whose structures are given.
4. Write the structural and condenses isomers when given the molecular formulas of organic ompounds.

MATERIALS

organic model kits

KEYED OBJECTIVES IN TEXT: 11-2, 11-3, 11-4, 11-5, 11-6, 11-7, 11-9

DISCUSSION OF EXPERIMENT

The hydrocarbons represent a group of organic compounds that are made of only two elements, carbon and hydrogen. In a compound, each carbon atom must always have four bonds, where each hydrogen atom has one bond. Each model of a hydrocarbon compound must follow this requirement.

The first compound you will construct is methane, CH_4. When you build the methane model, you will notice that it has a three-dimensional shape. To represent that structure on paper, mentally flatten out its shape, and write a structural formula. A structural formula always shows the individual bonds.

As the number of carbon atoms in a hydrocarbon increases, the writing of the structural formula becomes time consuming. For convenience, chemists use a shortened version called the condensed formula.   To convert a structural to a condensed formula, the number of hydrogen atoms are summarized and shown with a subscript after each carbon atom.

structural	condensed
H‑C‑H (with H above and below)	CH_4
H‑C‑C‑H (with H above and below each C)	CH_3CH_3

225

EXPERIMENT 30   ALKANES: STRUCTURES AND ISOMERS

A. STRUCTURE OF ALKANES

One family of hydrocarbons is called the **alkane family.** In the alkanes, the carbon-to-carbon bonds are always single bonds. Obtain an organic model kit to prepare the models of some alkanes. In the kit, the black wooden balls will represent carbon atoms, and the yellow wooden balls are the hydrogen atoms. The number of holes drilled in the wooden balls represent the number of covalent bonds that must be formed by that atom. The location of the holes represents the actual three-dimensional angles of the bonds.

A-1 Using the model kits, made a model of the alkane with 1 carbon atom called **methane.** The arrangement of the hydrogens about the carbon atom is called a tetrahedral shape. To represent the model on paper, mentally flatten out the molecule. Write its **structural formula** and its **condensed formula.**

A-2 Prepare models of alkanes that contain two(2) carbon atoms and three(3) carbon atoms. Write the structural and condensed formulas for each.

B. ISOMERS

When an alkane has four or more carbon atoms, it becomes possible to make more than one structure. One of the reasons for the vast array of organic compounds is the phenomenon of isomerism. Isomers have the same number of atoms, but the atoms are arranged in a different order. Isomers are present when a molecular formula can represent two or more different structural(or condensed) formulas.

B-1 Construct a model of an alkane with four(4) carbon atoms in a row. Write its structural and condensed formulas. This alkane represents a straight-chain structure and is called **butane.**

B-2 Using the model you prepared in B-1, remove the -CH_3 group from the end of the chain. Replace the hydrogen atom on the center carbon atom with the -CH_3 group. Fill in the rest of the chain with hydrogen atoms. This structure is called a **branched-chain** compound. Write its structural and condensed formulas. Typically the -CH_3 in the center is written above the chain.

B-2 There are three isomers for the molecular formula C_5H_{12}. Construct models of each isomer, and write the structural and condensed formulas. Write the condensed formula and name of each isomer.

C. CYCLOALKANES

The cycloalkanes are a kind of alkane that forms a cyclic or ring structure. There are no end carbon atoms since the carbon atoms are in a ring. To prepare models of cycloalkanes, you will need to use springs for the bonds between the carbon atoms.

The structural formula of a cycloalkane indicates all the carbon and hydrogen atoms and the condensed structure summarizes the hydrogen atoms for each carbon atom. Another type of notation called the **geometric structure** can be used to depict a cycloalkane. The geometric shape consists only of bonds that outline the geometric shape of the carbon atoms in the model. The actual atoms of carbon and hydrogen are not shown, but are understood. The geometric shape of the ring of three carbon atoms is a triangle while a square depicts a ring of four carbon atoms. An example using cyclobutane is shown below.

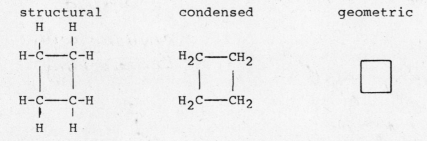

Draw the structure for cyclopropane, cyclobutane, and cyclopentane.

D. HALOALKANES

The haloalkanes are derivatives of alkanes in which a hydrogen atom has been replaced with a halogen atom.

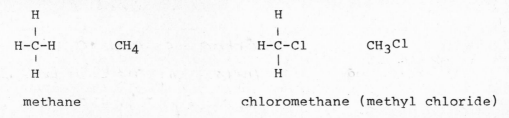

Construct models of haloalkanes using the wooden balls that represent the halogens: chlorine, green; bromine, orange,; iodine, violet. Draw the structural and condensed formulas.

alkanes - more random

$-\overset{|}{\underset{|}{C}}-\overset{|}{\underset{|}{C}}-$ (single)

alkenes - held rigidly

$\overset{}{\underset{}{C}}=\overset{}{\underset{}{C}}$ (double)

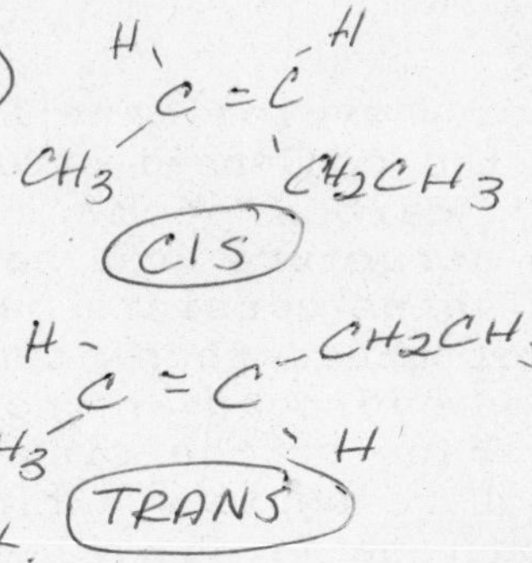

$\overset{H}{\underset{CH_3}{C}}=\overset{H}{\underset{CH_2CH_3}{C}}$ (CIS)

cis - same side
trans - across

$\overset{H}{\underset{CH_3}{C}}=\overset{CH_2CH_3}{\underset{H}{C}}$ (TRANS)

alkynes - no movement.

$-C\equiv C-$ (triple)

give multiple
bond lowest
substituent
(add -nding suffix)

$-\overset{|}{\underset{|}{C}}-C\equiv C-\overset{Cl}{\underset{|}{C}}-\overset{|}{\underset{|}{C}}-$

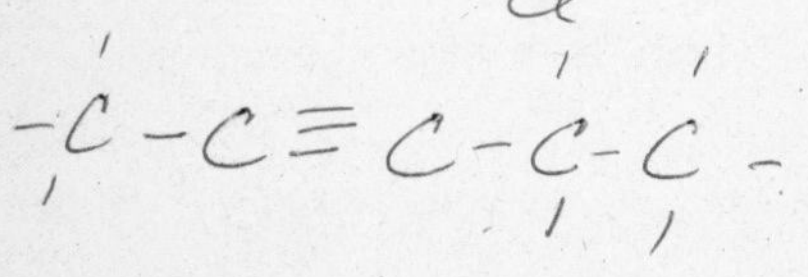

4 chloro-2-pentyne

aromatics

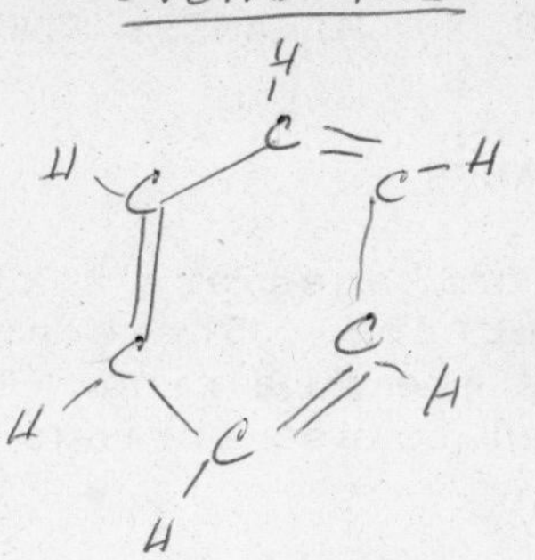

$\overset{}{\underset{}{C}}$

$H-\overset{}{\underset{}{C}} \quad \overset{4}{\underset{}{C}}=\overset{}{\underset{}{C}}-H$

$H-\overset{}{\underset{}{C}} \quad \overset{}{\underset{}{C}}$

$H \quad \overset{}{\underset{H}{C}} \quad H$

condenses to:

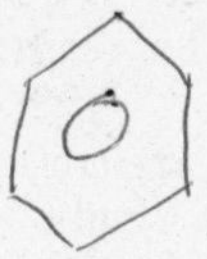

6 equal bonds
all the way
around.

- anything with
Benzene ring

Chlorobenzene
(no # needed)
when 1

Ortho - o next to
meta - m one C in between
para - p opposite sides.

O - dichlorobenzene
(1 benzene with
2 chlorines next each
other).

EXPERIMENT 33   ALDEHYDE AND KETONES

PURPOSE

1. Observe some physical characteristics of aldehydes and ketones.
2. Observe the oxidation of aldehydes using Benedict's reagent
3. Write the products of oxidation of aldehydes.

MATERIALS

test tubes and test tube rack
hot water bath
Benedict's reagent
organic compounds: acetaldehyde, benzaldehyde, vanillin,
                     5% glucose, cinnamaldehyde, camphor,
                     acetone (propanone)

Chemistry hand book

KEYED OBJECTIVES IN TEXT: 13-1, 13-6

DISCUSSION OF EXPERIMENT

Aldehydes and ketones contain the carbonyl functional group, - -.
In adehydes, the carbonyl group occurs at the end of the carbon
chain while ketones contain the carbonyl group on one of the carbon
atoms within the chain.

$$
\begin{array}{cccc}
O & O & O & O \\
\| & \| & \| & \| \\
CH_3CH & CH_3CH_2CH & CH_3CCH_3 & \text{⬡}-CH \\
\text{Acetaldehyde} & \text{Propionaldehyde} & \text{Acetone} & \text{Benzaldehyde} \\
 & & \text{(Propanone)} &
\end{array}
$$

OXIDATION OF ALDEHYDES

Aldehydes can be oxidized to form carboxylic acids. Ketones cannot
undergo further oxidation and do not react with oxidized agents.
The Benedict's reagent is a mild oxidizing agent that contains
cupric ion (Cu^{2+}) in a basic solution. When the aldehyde group is
oxidized, the cupric ion is reduced to cuprous ion (Cu^+) which
forms a reddish-orange precipitate, Cu_2O.

$$
\begin{array}{c}
O \\
\| \\
CH_3CH
\end{array}
+ \; 2\; Cu^{2+} \quad \xrightarrow{[O]} \quad
\begin{array}{c}
O \\
\| \\
CH_3COH
\end{array}
+ \quad Cu_2O(s)
$$

$$\text{blue} \qquad\qquad\qquad\qquad\qquad \text{red-orange}$$

$$
\begin{array}{c}
O \\
\| \\
CH_3CCH_3
\end{array}
+ \; 2 \; Cu^{2+} \quad \xrightarrow{[O]} \quad \text{No reaction}
$$

EXPERIMENT 33   ALDEHYDES AND KETONES

LABORATORY ACTIVITIES

A.   MODELS OF ALDEHYDES AND KETONES

Obtain an organic model kit and construct models of formaldehyde, acetaldehyde, acetone, and butanone(ethyl methyl ketone).

B.   PHYSICAL PROPERTIES

Using a chemistry handbook or Merck's Index, write the structure of acetone, camphor, cinnamaldehyde, benzaldehyde and vanillin. Carefully detect the odor of each. Circle the functional group in each structure and name the group. Record the melting and boiling point of each. Indicate some commercial or medicinal uses for the compound.

C.   OXIDATION OF ALDEHYDES AND KETONES USING BENEDICT'S REAGENT

C-1   Write the structure of each compound listed in the laboratory record.

C-2   Place 2 mL of organic liquid in separate test tubes. Label each. Add 5 mL of Benedict't reagent to each sample. Place the test tubes in a boiling water bath for 5-10 minutes only. **CAUTION: ACETONE IS FLAMMABLE.** Record and change in color.

C-3   Indicate whether the compound was oxidized. A change in color from blue to any shade of green to a red-orange indicates that oxidation has occurred. Write the formula of the product.

Glucose is an example of a carbohydrate that contains an aldehyde group. The oxidation of glucose with Benedict's reagent is a typical clinical test for detecting glucose in the urine.

$$
\begin{array}{c}
O \\
\parallel \\
CH \\
| \\
HCOH \\
| \\
HOCH \\
| \\
HCOH \\
| \\
HCOH \\
| \\
CH_2OH
\end{array}
$$

D-GLUCOSE

250

EXPERIMENT 34   CARBOXYLIC ACIDS AND ESTERS

PURPOSE

1. Prepare models of carboxylic acids and esters.
2. Determine the solubility of carboxylic acids and their salts.
3. Write structures of the prepared esters.
4. Observe the hydrolysis of an ester.

MATERIALS

organic model kits
carboxylic acids: acetic acid, propionic acid, tartaric acid,
                   benzoic acid, salicylic acid
alcohols: methanol, isoamyl alcohol, ethanol, benzyl alcohol
           1-propanol, octanol

test tubes and test tube rack	6 M NaOH
litmus paper	beakers
con. H_2SO_4	10% $NaHCO_3$
hot water bath	glacial acetic acid
reflux apparatus	$MgSO_4$ anhydrous
separatory funnel	distillation apparatus
cotton	methyl salicylate

KEYED OBJECTIVES IN TEXT: 13-5, 13-6, 13-7, 13-8, 13-9, 13-10

DISCUSSION OF EXPERIMENT

Carboxylic acids have one or more carboxyl ($-\overset{\overset{O}{\|}}{C}OH$) groups. The
polarity of the carboxyl group makes low molecular weight acids,
and diacids with two or more carboxyl groups soluble in water.

$$CH_3\overset{\overset{O}{\|}}{C}OH \qquad CH_3CH_2\overset{\overset{O}{\|}}{C}OH \qquad \bigcirc\!\!\!-\overset{\overset{O}{\|}}{C}OH \qquad HO\overset{\overset{O}{\|}}{C}-\overset{\overset{OH}{|}}{C}H-\overset{\overset{OH}{|}}{C}H-\overset{\overset{O}{\|}}{C}OH$$

Acetic acid   Propionic acid   Benzoic acid   Tartaric acid

They act as weak acids in aqueous solution when a small percent of
the carboxyl groups dissociate.

$$CH_3\overset{\overset{O}{\|}}{C}OH \quad \rightleftharpoons \quad CH_3\overset{\overset{O}{\|}}{C}O^- \quad + \quad H^+$$

Acetic acid   Acetate ion

EXPERIMENT 34   CARBOXYLIC ACIDS AND ESTERS

NEUTRALIZATION OF CARBOXYLIC ACIDS

Carboxylic acid are neutralized by bases such as sodium hydroxide to form salts. These salts are usually soluble in water.

$$CH_3\overset{\overset{\displaystyle O}{\|}}{C}OH \quad + \quad NaOH \quad \longrightarrow \quad CH_3\overset{\overset{\displaystyle O}{\|}}{C}O^-\ Na^+ \quad + \quad H_2O$$

Acetic acid   Sodium acetate

ESTERS

An important reaction of carboxylic acids is their esterification with alcohols. Many esters have fragrant odors.

$$CH_3\overset{\overset{\displaystyle O}{\|}}{C}O(CH_2)_7CH_3 \qquad CH_3\overset{\overset{\displaystyle O}{\|}}{C}OCH_2CH_2\underset{\underset{\displaystyle CH_3}{|}}{C}HCH_3$$

Octyl acetate   Isoamyl acetate   Methyl salicylate
(orange)   (pear)   (wintergreen)

Esterification occurs with a carboxylic acid and an alcohol in the presence of an acid.

$$CH_3\overset{\overset{\displaystyle O}{\|}}{C}OH \quad + \quad HOCH_2CH_3 \quad \xrightarrow{H^+} \quad CH_3\overset{\overset{\displaystyle O}{\|}}{C}O\text{-}CH_2CH_3 \quad + \quad H_2O$$

Acetic acid   Ethanol   Ethyl acetate

The saponification of an ester occurs when it is hydrolyzed in the presence of a base. The products are the salt of the carboxylic acid and the alcohol.

$$CH_3\overset{\overset{\displaystyle O}{\|}}{C}OCH_2CH_3 \quad + \quad NaOH \quad \longrightarrow \quad CH_3\overset{\overset{\displaystyle O}{\|}}{C}O^-Na^+ \quad + \quad CH_3CH_2OH \quad + \quad H_2O$$

Ethyl acetate   Sodium acetate   Ethanol

EXPERIMENT 34   CARBOXYLIC ACIDS AND ESTERS

LABORATORY ACTIVITIES

A. SOLUBILITY OF CARBOXYLIC ACIDS

A-1 Write the structures of the carboxylic acids that will be used in this experiment.

        acetic acid, propionic acid, tartaric acid, benzoic acid

A-2 Place 4 mL of water in four separate test tubes. Add 5 drops of the acid (or the amount on the tip of a microspatula). Stopper (cork) and shake. If the acid dissolves, add 5 more drops (or another scoop). Record your observations as soluble, slightly soluble, or insoluble.

A-3 Place the test tubes in a hot water bath and heat for 5 minutes. Record your observations (soluble, insoluble).

A-4 Test the pH of each sample by using a glass stirring rod to place a drop of the solution on the piece of pH paper. Report the pH of the solution.

A-5 Add 10-15 drops of 6 M NaOH to each sample. Stir and record your observations (soluble or insoluble)

A-6 Write the formulas of the sodium salts of each carboxylic acid.

A-7 Write the equation for the neutralization of the carboxylic acids.

B.   PREPARATION OF SOME ESTERS

B-1 Using the organic model kit, prepare models of acetic acid and ethanol. Remove the components of water and join the remaining portions to form a model of the ester. Write the equation for this reaction and name the ester.

B-2 Place the following combinations in separate test tubes:

     (1) 2 mL glacial acetic and 2 mL isoamyl alcohol
     (2) 2 mL methanol and 1 g salicylic acid

     CAUTION: USE CARE IN DISPENSING GLACIAL ACETIC ACID.   IT CAN CAUSE BURNS AND BLISTERS ON THE SKIN.

     Add 5 drops of concentrated H_2SO_4 to each test tube.
        CAUTION: USE CARE IN THE ADDITION OF SULFURIC ACID. WEAR GOGGLES!

Place the test tubes in a hot water bath(70-80°C) for 10 minutes. CAUTIOUSLY note the odor of each. If no odor is detectable, pour the contents of the test tube into a beaker containing 20 mL of warm water. Identify odors such as pear, banana, or oil of wintergreen.

Write the equations for the reactions and name the esters.

C.   SAPONIFICATION OF AN ESTER

C-1   Place 10 drops of methyl salicylate and 2 mL of water in a test tube. Record the appearance and odor of the ester.

C-2   Add 3 mL of 6 M NaOH. Place the test tube in a boiling hot water bath for 10-15. Record any changes in the odor and appearance of the ester.

C-3   Write the equation for the saponification reaction.

D.   OPTIONAL: PREPARATION OF AN ESTER

If refluxing equipment is available, an ester may be prepared and isolated.

D-1   Set up a reflux apparatus using a 200-mL round-bottom flask. See Figure 34-1.

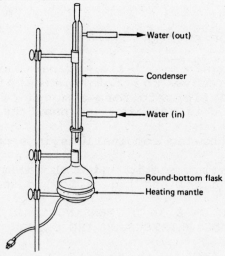

Figure 34-1   Reflux apparatus.

EXPERIMENT 34   CARBOXYLIC ACIDS AND ESTERS

D-2   Place the alcohol and acid from one of the following pairs in the flask and mix.

   alcohol   carboxylic acid

   isoamyl alcohol(30 mL) and glacial acetic acid(30 mL) CAREFUL

   benzyl alcohol (25 mL) and glacial acetic acid (30 mL)

   methanol (50 mL)   and salicylic acid (10 g)

   octanol (25 mL) and   glacial acetic acid (30 mL)

D-3   SLOWLY add 5 mL of concentrated H_2SO_4. GOGGLES. Swirl the flask to mix the reactants and add some boiling chips to the solution. Attach the reflux condenser vertically to the flask. Turn on the water so the water flows into the lower tubing leading to the condenser and out the upper tubing. Reflux the mixture for 1 to 2 hours. Remove the reflux condenser and let the solution cool by immersing the reaction flask in a cold water bath.

D-4   Pour the contents of the flask into 100 mL of cold water contained in a large beaker. Pour this mixture into a separatory funnel. You should see two layers in the funnel. Be sure you know which layer is the ester layer. Separate the ester layer from the water layer and discard the water layer.

D-5   Wash the ester layer with 50 mL of 10% $NaHCO_3$, swirling the mixture in the funnel. Do not stopper the funnel until you see that CO_2 gas is no longer being released.

   CAUTION: Stopper, then invert the funnel in your hand holding the stopper. Immediately open the stopcock to release the pressure which will build up inside the funnel. Do not let the funnel point at anyone. Close the stopcock and cautiously shake the funnel, stopping often to invert and release pressure.

D-6   Place the funnel in an iron ring and remove the stopper. Allow the two layers to separate. Open the stopcock and allow the lower layer to flow into an Erlenmeyer flask. Pour the upper layer into a second flask. Discard the water layer. Add a small amount of anhydrous magnesium sulfate to dry the ester. The solution should become clear. If not, add some more anhydrous magnesium sulfate and let the solution stand. See Figure 34-2.

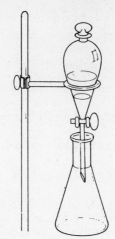

Figure 34-2   Separatory funnel setting in an iron ring while layers separate.

D-7   Set up a distillation apparatus using a 100-mL distillation flask. See Figure 34-3. Be sure that all pieces are dry. Filter the solution into the distilling flask by pouring it through a funnel fitted with a cotton plug. Add several boiling chips to the distilling flask and distill the ester with low heat, slowly. Since the vapor pressure of these esters are high, collect your products in a receiving flask set in ice. DO NOT OVERHEAT OR LET THE DISTILLATION FLASK BECOME DRY.   Note the temperature range of product distillation.

D-8   Weigh or measure the volume of your product. Place the product in a shell vial and stopper. Attach a label with your name, the name of the product, its mass and temperature range during distillation.

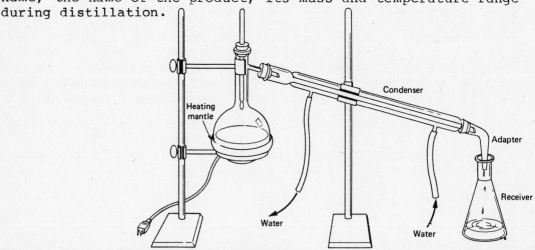

Figure 34-3   Distillation apparatus.

EXPERIMENT 34   CARBOXYLIC ACIDS AND ESTERS   NAME_____
                                                SECTION_____
LABORATORY RECORD   DATE_____

A. SOLUBILITY OF CARBOXYLIC ACIDS

	Acetic acid	Propionic acid	Tartaric acid	Benzoic acid
A-1 Structure				
A-2 Solubility (cold water)				
A-3 Solubility (hot water)				
A-4 pH				
A-5 Solubility in NaOH				
A-6 Structure of sodium salt				

A-7 Equations for neutralization

acetic acid _____

propionic acid _____

tartaric acid _____

benzoic acid _____

261

B.   PREPARATION OF SOME ESTERS

B-1 Esterification equation for acetic acid and ethanol

B-2 Structure of reactants	Structure and name of ester	Fragrance
(1) $CH_3\overset{\overset{\displaystyle O}{\|}}{C}OH$ $CH_3CHCH_2CH_2OH$ Acetic isoamyl acid alcohol		
(2) CH_3OH Salicylic methanol		

C. SAPONIFICATION OF AN ESTER

C-1 Appearance and odor of ester

C-2 Appearance and odor of products after saponification

C-3 Equation for the saponification of the ester in NaOH

EXPERIMENT 34   CARBOXYLIC ACIDS AND ESTERS

D. OPTIONAL: PREPARATION OF AN ESTER

Acid and alcohol used _____

Ester formed   _____

Fragrance   _____

Distillation range, ^{o}C _____

Actual boiling point. ^{o}C_____

Volume or mass obtained _____

EXPERIMENT 35   PREPARATION OF ASPIRIN

PURPOSE

1. Use laboratory methods to prepare aspirin.
2. Purify the crude aspirin sample.
3. Test for aspirin purity.

MATERIALS

salicylic acid	1% $FeCl_3$
large test tube	test tubes (3)
thermometer	5- or 10-mL graduated cylinder
stirring rod	commerical aspirin
acetic anhydride	funnel and filter paper
conc. (85%) H_3PO_4	
melting point apparatus	
ice water	
watch glass	
Buchner funnel setup	

KEYED OBJECTIVES IN TEXT: 13-8

Aspirin is a fever reducer and pain reliever found in various commerical products. In such products, the aspirin is bound together with a starch binder. Aspirin(acetylsalicyclic acid) can be prepared by esterifying acetic acid with the alcohol group on salicylic acid. However, this is a slow reaction. The ester forms more rapidly when acetic anhydride is used to provide the acetyl group.

Salicylic acid   Acetic anhydride   Aspirin   Acetic acid

The purity of the aspirin can be determined by adding ferric chloride. If there is unreacted salicylic acid present, the free phenolic group reacts to tive a violet color.

265

EXPERIMENT 35   PREPARATION OF ASPIRIN

LABORATORY ACTIVITIES

SYNTHESIS OF ASPIRIN

1. Prepare a hot water bath using a 400-mL beaker. Place 2.0 g
    of salicylic acid in a large test tube. Add 5-mL acetic
    anhydride to the test tube.

    CAUTION: ACETIC ANHYDRIDE IS IRRITATING TO THE MEMBRANES OF
    THE NOSE AND SINUS.   WEAR GOGGLES.

2. Slowly add 5 drops of concentrated (85%) phosphoric acid.
    Plce the test tube in the hot water bath and heat until all of
    the solid dissolves. Remove the test tube and let the
    contents cool.

3. CAUTIOUSLY add 5 mL of water to the test tube. KEEP YOU FACE
    AWAY FROM THE MOUTH OF THE TEST TUBE. The water decomposes
    unreacted acetic anhydride which can be very reactive.   The
    vapors emitted may contain acetic acid fumes which are very
    irritating.

4. If no crystals appear when the test tube contents have cooled
    to room temperature, place the test tube in ice water and
    scratch the sides with a stirring rod. Crystals of aspirin
    should form. When crystallization is complete, collect the
    crystals by pouring the contents of the test tube into a
    Buchner funnel using suction filtration. (See Figure 35-1)
    Wash out the test tube with 5 or 10 mL of ice water. Wash the
    crystals on the filter paper several times with 5 mL ice
    water. Continue to draw air through the funnel to remove as
    much water as possible and to dry the crystals. This is
    aspirin, but it not yet purified and is considered to be a
    crude product. Place your aspirin product onto a watch glass
    to dry.

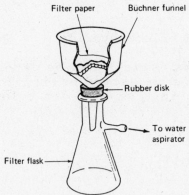

Figure 35-1   Apparatus for suction filtration.

EXPERIMENT 35   PREPARATION OF ASPIRIN

A-1   When the crude product is dried, determine the mass of the product and record. Remove a small sample of the crude product and store for later testing.

A-2   Calculate the percentage yield of crude aspirin product. The theorectical yield of aspirin from 2.0 g salicylic acid would be 2.61 g.

$$\text{percent yield} = \frac{\text{mass of product}}{2.61 \text{ g}} \times 100\%$$

A-3   OPTIONAL: The major impurity in the aspirin product is unreacted salicylic acid. To purify your crude aspirin product, place the crystals in a small beaker. Add 15 mL ethanol. Add 30 mL warm water to the beaker dissolving all the aspirin. Let the solution cool. You may cool further by placing the beaker in ice. Crystals should form which are collected the same way that you did before. Let the aspirin dry. Weigh the dried purified aspirin product. Record.

A-4   Calculate the percent yield of the purified aspirin product.

A-5   Obtain a melting point apparatus and determine the melting point of a few crystals of the crude and the purified product.

A-6   Obtain a chemistry handbook and look up the actual melting point.

B.   DETERMINING THE PURITY OF ASPIRIN

B-1   Place 1 mL of methanol in four separate test tubes. Add a few crystals of each of the following:

     (1) salicylic acid
     (2) crude product
     (3) purified product
     (4) commerical aspirin

Add 1 drop of 1% ferric chloride solution to each test tube. Record the color.

Place the crude or purified product in a shell vial. Stopper, attach a label with your name, mass and melting point, and turn in to your instructor.

EXPERIMENT 35   PREPARATION OF ASPIRIN   NAME_____
                                           SECTION_____
LABORATORY RECORD   DATE_____

A.   PREPARATION OF ASPIRIN

                        CRUDE   PURIFIED
                        PRODUCT   PRODUCT

MASS   _____   _____

PERCENT YIELD   _____   _____

MELTING POINT   _____   _____

Handbook value for melting point _____

B.   PURITY

COMPOUND   COLOR WITH FeCl$_3$ IMPURITIES(YES/NO)

Salicylic acid   _____   _____

Crude aspirin   _____   _____

Purified aspirin   _____   _____

Commerical aspirin_____   _____

QUESTIONS

Which of the samples tested would be considered to have salicylic
acid impurities?   Why

What of the samples, if any, would be considered pure?

EXPERIMENT 36   AMINES AND AMIDES

PURPOSE

1. Write the structure and names of amines.
2. Classify amines as primary, secondary or tertiary.
3. Writean equation for the formation of an amine salt.
4. Write equations for amidation reactions.
5. Write equations for the hydrolysis of amides.

MATERIALS

organic model kits   Buchner funnel
test tubes   Short stem funnel
triethylamine   filter paper
benzamide   pH or litmus paper
methyl benzoate   impure acetanilide
conc. ammonia
6 M HCl
10% NaOH

KEYED OBJECTIVES IN TEXT: 14-1, 14-2, 14-4, 14-5, 14-6, 14-7

DISCUSSION OF EXPERIMENT

Amines are a class of compounds that are derivatives of ammonia.
In amines, organic groups may replace one, two or three of the
hydrogens of ammonia to give primary, secondary or tertiary amines.

$$NH_3 \qquad CH_3NH_2 \qquad CH_3NHCH_3 \qquad CH_3\overset{\displaystyle CH_3}{\underset{\displaystyle |}{N}}CH_3$$

Ammonia   Methylamine   Dimethylamine   Trimethylamine
             (primary)   (secondary)   (tertiary)

AMINES IN WATER

Amines act as weak bases in water due to the ability of the
nitrogen atom to attract protons to its unshared pair of electrons.

$$CH_3NH_2 \quad + \quad H_2O \longrightarrow CH_3NH_3^+ \quad + \quad OH^-$$

Methylamine   Methylammonium ion

NEUTRALIZATION OF AMINES

In an acidic solution, an amine is neutralized to give its
salt which is soluble in water.

$$CH_3NH_2 \quad + \quad HCl \longrightarrow CH_3NH_3^+ \; Cl^-$$

Methylamine   Methylammonium chloride

271

FORMATION OF AMIDES

When amines react with carboxylic acids, amides are produced.

$$CH_3\overset{\overset{\displaystyle O}{\|}}{C}OH + NH_3 \longrightarrow CH_3\overset{\overset{\displaystyle O}{\|}}{C}NH_2 + H_2O$$

Acetic acid   Ammonia   Acetamide

$$CH_3\overset{\overset{\displaystyle O}{\|}}{C}OH + CH_3NH_2 \longrightarrow CH_3\overset{\overset{\displaystyle O}{\|}}{C}NHCH_3 + H_2O$$

Acetic acid   Methylamine   N-Methylacetamide

LABORATORY ACTIVITIES

A.   STRUCTURE AND CLASSIFICATION OF AMINES

A-1 Use a model kit to construct models of ammonia and the
     following amines:

     methylamine, ethylamine, dimethylamine, trimethylamine,
     dimethylethylamine

     Write the condensed structure of each model in the laboratory
     record.

A-2 Classify each amine as primary (1^O), secondary (2^O) or
     tertiary (3°).

B.   IONIZATION OF AMINES IN WATER

     WORK IN THE HOOD.   THE VAPORS OF AMINES ARE IRRITATING TO NOSE
     AND SINUS.

B-1 Place 5 mL of water in a test tube.   Add 10 drops of
     triethylamine. Record its solubility in water.

B-2 CAUTIOUSLY note the odor.

B-3 Use a stirring rod to place a drop of the solution on pH or
     litmus paper. Record the results.

B-4 Write the equation for the ionization of the amine in water.

B-5 Add 6 M HCl dropwise to the amine solution until the solution
     is acidic with litmus paper.   Record any changes in the
     solubility of the amine.

B-6 Cautiously note its odor.

B-7 Write the neutralization equation for the reaction of the amine with HCl.

C. STRUCTURE OF AMIDES

Use the organic model kits to form models of acetic acid, benzoic acid and two ammonia molecules. Combine a portion of each acid with ammonia to form an amide and water. Write the equation for each amidation and the name of the products.

D. FORMATION OF AN AMIDE

THIS MAY BE CARRIED OUT BY YOUR INSTRUCTOR AS A DEMONSTRATION

The amide, benzamide, may be produce by combining methyl benzoate and concentrated ammonia. USE CAREFULLY!

$$\underset{\text{Methyl benzoate}}{\text{C}_6\text{H}_5-\text{COCH}_3} \quad + \quad \underset{\text{ammonia}}{\text{NH}_3} \quad \longrightarrow \quad \underset{\text{benzamide}}{\text{C}_6\text{H}_5-\text{CNH}_2} \quad + \quad \underset{\text{methanol}}{\text{CH}_3\text{OH}}$$

Place 10 mL of methyl benzoate and 30 mL of concentrated ammonia in a 125-mL Erlenmeyer flask. Stopper and set in the hood. Do not disturb for a few days. At the next laboratory, look for the formation of crystals of benzamide. If crystals are present, filter and discard the filtrate. CAUTION: THE MIXTURE STILL CONTAINS CONCENTRATED AMMONIA.

When the crystals are dry, take a melting point and turn in your product. Use a handbook to determine the actual melting point of benzamide. Record.

E. HYDROLYSIS OF AN AMIDE

An amide is hydrolyzed by either an acid or a base.

$$\underset{}{\text{CH}_3\text{CNH}_2} \quad + \quad \text{HCl} \quad \longrightarrow \quad \text{CH}_3\text{COH} \quad + \quad \text{NH}_4\text{Cl}$$

$$\underset{}{\text{CH}_3\text{CNH}_2} \quad + \quad \text{NaOH} \quad \longrightarrow \quad \text{CH}_3\text{CO}^-\text{Na}^+ \quad \text{NH}_3$$

E-1 Place 0.5 g benzamide in a test tube. Add 5 mL of 10 % NaOH. CAUTIOUSLY note the odor of the mixture.

E-2 Place the test tube in a boiling water bath and heat gently. Hold a piece of moistened litmus paper over the mouth of the test tube. Record the color of the litmus paper.

E-3 CAUTIOUSLY note any odor from the test tube. Record.

E-4 Write the equation for the hydrolysis of benzamide by NaOH.

F.   ISOLATION OF ACETANILIDE

The amide acetanilide shows a tenfold increase in solubility from $25^{\circ}C$ to $100^{\circ}C$. This difference in solubility can be used to isolate and purity acetanilide from an impure sample.

F-1   Weigh a 250-mL beaker. Add approximately 2 g of impure acetanilide.   Reweigh the beaker and contents. Record.

F-2 Calculate the mass of the impure sample.

F-3   Add 50 mL of water to the beaker and heat the mixture until no more solid material appears to dissolve. While the mixture is heating, heat another beaker of water for use in the filtration.

Set up a short-stem funnel fitted with filter paper in an iron ring. When you are ready to filter your impure sample, pour some of the hot water through the funnel to warm the glass. Filter the warm sample of impure acetanilide into a clean beaker or flask. Rinse the funnel with hot water before crystals form on it. Place the beaker or flask containing the warm filtrate in an ice bath.

As the filtrate cools, crystals of acetanilide should form. Collect the crystals by using the Buchner funnel and suction filtration apparatus. Dry the crystals on a clean filter paper or watch glass.

Determine the mass of the dried product and record.

F-4 Calculate the percent recovery of the pure product.

$$\% \text{ recovery} = \frac{\text{mass of purified product}}{\text{mass of impure sample}} \times 100\%$$

F-5 Use a melting point apparatus to determine the melting point of your product.

F-6 Use a chemistry handbook to determine the actual melting point.   Record.

Place the purified product in a shall vial, stopper, label with your name, % recovery and melting point of the product. Turn in to your instructor.

EXPERIMENT 36 AMINES AND AMIDES

NAME______________

SECTION______________

LABORATORY RECORD

DATE______________

A. STRUCTURE AND CLASSIFICATION OF AMINES

COMPOUND	STRUCTURE	CLASSIFICATION
Methylamine	______________	________
Ethylamine	______________	________
Dimethylamine	______________	________
Trimethylamine	______________	________
Dimethylethylamine	______________	________

B. IONIZATION OF AMINES

B-1 Solubility of triethylamine in water ______________

B-2 Odor ______________

B-3 Acidic or basic? ______________

B-4 Equation for ionization in water

____

B-5 Solubility after adding HCl ______________

B-6 Odor after adding HCl ______________

B-7 Equation for neutralization with HCl

____

C.   STRUCTURE OF AMIDES

   Equations for amidation:

   acetic acid and ammonia

   benzoic acid and ammonia

D.   FORMATION OF AN AMIDE

   Melting point of product   _____

   Handbook value   _____

E.   HYDROLYSIS OF AN AMIDE

E-1 Odor of benzamide   _____

E-2 Color of pH or litmus paper   _____

E-3 Odor after hydrolysis   _____

E-4 Equation for the NaOH hydrolysis of benzamide

F.   ISOLATION OF ACETANILIDE

F-1 Mass of beaker and impure sample _____

   Mass of beaker   _____

F-2 Mass of impure sample   _____

F-3 Mass of purified product   _____

F-4 Percent recovery   _____

F-5 Melting point of acetanilide   _____

F-6 Handbook value   _____

EXPERIMENT 37   CARBOHYDRATES

PURPOSE

1. Use chemical tests to identify some physical and chemical characteristics of typical carbohydrates.
2. Use chemical test to differentiate between monosaccharides, disaccharides, and polysaccharides.
3. Identify an unknown carbohydrate.

MATERIALS
1% solutions: glucose, fructose, sucrose, lactose,
               maltose, starch, glycogen
test tubes
boiling water bath
reagents: Benedict's, iodine, Seliwanoff's, Barfoed's
Baker's yeast
6M HCl, 6M NaOH
fermenation tubes (or test tubes with smaller tubes to fit inside)
foods: sugars (refined, brown,"natural", powdered)
        syrups (corn, maple, fruit)
        honey, saccharin, cereals
        bread, aspirin

KEYED OBJECTIVES IN TEXT: 15-1, 15-2, 15-3, 15-4, 15-5, 15-6

DISCUSSION OF EXPERIMENT

Carbohydrates are the major compounds in our diet that provide energy to run chemical reactions in the body and make us move. Carbohydrates are polyhydroxy compounds that have either a keto group or an aldehyde group. Glucose is a typical monosaccharide while maltose is a typical disaccharide. Glucose exists almost entirely in the closed-chain hemiacetal form. In maltose, two glucose units react to form an acetal.

α –D-Glucose   β –Maltose

BENEDICT'S TEST FOR REDUCING SUGARS

All of the monosaccharides and most of the disaccharides are easily oxidized. While the monosaccharide exists as a hemiacetal most of the time, there is always a small amount that reverts to the open-chain form which contains the aldehyde group. Both aldoses and ketoses react with Benedict's reagent.

EXPERIMENT 37   CARBOHYDRATES

By being oxidized, the sugars cause the reduction of some
other substance (thus the name <u>reducing sugars</u>).   In this
experiment, you will use Benedict's reagent to test for the
presence of reducing sugars among the mono- and disaccharides.
When the cupric ion, Cu^{2+}, reacts with a reducing sugar, a red
precipitate of cuprous oxide, Cu_2O,(s) is formed.

$$\text{sugar-CH} \quad + \quad 2\ Cu^{2+} \quad \longrightarrow \quad \text{sugar-COH} \quad + \quad Cu_2O(s)$$

reducing sugar   blue   oxidized sugar   red-orange

Benedict's test is also a useful clinical test in which the
reagents are present in a tablet that is dropped into a test tube
containing water and a few drops of urine.   The formation of
cuprous oxide can be associated with different colors and with
different concentrations of glucose in a sample.

Sucrose, a disaccharide, does not react with Benedict's
reagent. In sucrose, the monosaccharides, glucose and fructose,
are bonded through the oxygen atoms of the hemiacetal parts of the
molecules. Therefore, there is no hemiacetal that can revert to
the open-chain form that would cause the reduction of the cupric
ion.

BARFOED'S TEST

Barfoed's test is similar to the Benedict's test, but it can
be used to distinguish between monosaccharides and disaccharides.
The cupric ion in the acidic reagent is oxidized much more rapidly
by monosaccharides to produce the red cuprous oxide(Cu_2O).   This
difference in rate of reaction is used to identify the presence of
monosaccharides.

SELIWANOFF'S TEST

Seliwanoff's test can be used to distinguish between
ketohexoses and aldohexoses. When ketohexoses are present, a deep
red color is formed rapidly. Aldohexoses give a light pink color
which develops more slowly.

IODINE TEST FOR POLYSACCHARIDES

Starch, a polysaccharide, contains two polymers, amylopectin
and amylose. Both consist of many glucose units with amylose being
a straight-chain polymer and amylopectin is a branched-chain
polymer.   When iodine is added to amylose, a deep, blue-black
color appears. Other polysaccharides give red to brown colors with
iodine. The monosaccharides and disaccharides are not reactive.

278

EXPERIMENT 37   CARBOHYDRATES

HYDROLYSIS OF DISACCHARIDES AND POLYSACCHARIDES

Disaccharides hydrolyze in the presence of an acid to give their individual monosaccharides. Polysaccharides such as amylose in starch can be hydrolyzed to smaller polysaccharides (dextrins) and to maltose. Complete hydrolysis produces many glucose molecules. This same type of hydrolysis of starch is also catalyzed by salivary and pancreatic enzymes during the digestion process.

FERMENTATION

Some sugars will ferment in the presence of yeast, which contains the enzyme zymase. The products of fermentation are ethyl alcohol and carbon dioxide. The formation of bubbles of carbon dioxide is used as confirmation of the fermentation process.

LABORATORY ACTIVITIES

A. STRUCTURAL FORMULAS FOR CARBOHYDRATES

Monosaccharides   Draw the Fischer projection and the Haworth cyclic formulas(α, β) for glucose and fructose.

Disaccharides   Draw the Haworth cyclic formulas for lactose, maltose and sucrose. Indicate the glycosidic bond.

Polysaccharides   Draw representative formulas of amylose and glycogen indicating the typical glycosidic bonds in each type of compound.

B. BENEDICT'S TEST FOR REDUCING SUGARS

Place 1 mL (20 drops) of each 1% carbohydrate solution in separate test tubes. Place 1 mL (20 drops) of water in another test tube as a control. Add 5 mL of Benedict's reagent to each test tube. Label and place all test tubes in a boiling water bath for 5 min. DO NOT OVERHEAT. Remove the test tubes and make your observations. The formation of a greenish-orange to reddish-orange color due to formation of the cuprous oxide precipitate indicates the presence of a reducing sugar. Record your results. Identify the compounds as reducing sugars or nonreducing sugars.

EXPERIMENT 37   CARBOHYDRATES

C. BARFOED'S TEST FOR MONOSACCHARIDES

Place 1 mL (20 drops) of each 1% carbohydrate solutions in separate test tubes. Prepare a control with 1 mL water. Add 4 mL Barfoed's reagent. **CAUTION: Barfoed's reagent contains acetic acid. WEAR GOGGLES.**

Place the test tubes in a boiling water bath for 10 min. Remove the test tubes and cool in cold water. Record your observations. A precipitate of cuprous oxide which may be green to red, indicates that a monosaccharide is present. Record your results and conclusions. Identify the compounds that are monosaccharides.

D. SELIWANOFF'S TEST FOR KETOHEXOSES

Place 1 mL of the 1% carbohydrate solution in separate test tubes. Prepare a control with 1 mL of water. Add 5 mL of Seliwanoff's reagent to each test tube. **CAUTION: WEAR GOGGLES. Seliwanoff's reagent contains HCl.**

Place the test tubes in a boiling hot water bath. After 1 min, observe the colors in the test tubes. Record your results. The rapid formation of a bright-red product indicates that a ketohexose is present. Identify the compounds that are ketohexoses.

E. IODINE TEST FOR POLYSACCHARIDES

Place 2 mL of each carbohydrate solution in separate test tubes. Prepare a control with 2 mL of water. Add 1 drop of iodine solution to each sample. Record your results. A dark color (red, brown, blue-black) is a positive test for a polysaccharide. Indicate the compounds that are polysaccharides.

F. HYDROLYSIS OF DI- AND POLYSACCHARIDES

Place 5 mL of sucrose and starch in separate test tubes. Add 1 mL of 6M HCl to each sample. Place the test tubes in a boiling water bath and heat for 20 min. Remove the test tubes and let them cool. Neutralize each sample with 6M NaOH (about 20 drops) until a drop of the mixture turns litmus paper blue.

F-1   Test 3 mL of each mixture with Benedict's reagent. Record your results and compare with the previous results of sucrose and starch with Benedict's reagent. Account for the differences observed.

F-2   Test the remainder of each mixture with iodine solution. Record your results and compare them with the previous results of sucrose and starch with iodine. Account for any differences in test results.

G. FERMENTATION TEST

Fill fermentation tubes with samples of the different 1% carbohydrate solutions. See Figure 37-1. Add 1 g yeast to each tube and mix well. Set the tubes aside. At the end of the laboratory period, and again at the next laboratory period, observe any changes in the solutions in the fermentation tubes. Record your observations.

Figure 37-1   Fermentation tube.

If fermentation tubes are not available, place the mixtures in 5 test tubes. Place smaller test tubes upside down in the larger test tubes. Place your hand firmly over the mouth of the test tube and invert. When the small test tube inside has completely filled with the mixture, return the larger test tube to an upright position. See Figure 37-2.

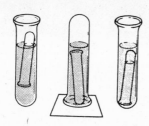

Figure 37-2   Test tubes used as fermentation tubes.

Set the tubes aside. At the end of the laboratory period, and again at the next laboratory period, look for bubbles of carbon dioxide in the fermentation tubes.   See Figure 37-3. Record your observations.

Figure 37-3   Fermentation tubes with CO_2 bubble.

H.   TESTING AN UNKNOWN CARBOHYDRATE

Use the carbohydrate test to identify an unknown carbohydrate sample as a mono-, di-, or polysaccharide. The possible compounds are glucose, fructose, sucrose, maltose, amylose(starch) and glycogen.   Record the results for each test and your conclusions.

I.   TESTING FOODS FOR CARBOHYDRATES

Obtain 3 samples of the food products available in the laboratory or brought from home. Perform the various tests for carbohydrates and identify as many kinds of carbohydrates as you can in each sample.

CARBOHYDRATES   NAME_____
                                                 SECTION_____
LABORATORY RECORD   DATE_____

A. STRUCTURAL FORMULAS FOR CARBOHYDRATES

Structures

MONOSACCHARIDES	Fischer	Haworth
glucose		
fructose		

DISACCHARIDES (Haworth structures only)

lactose	sucrose
maltose	
POLYSACCHARIDES amylose	glycogen

283

B. BENEDICT'S TEST FOR REDUCING SUGARS

Compound	Color	Reducing Sugar (yes/no)
water	_____	_____
glucose	_____	_____
fructose	_____	_____
sucrose	_____	_____
lactose	_____	_____
maltose	_____	_____
starch	_____	_____
glycogen	_____	_____

C. BARFOED'S TEST FOR MONOSACCHARIDES

Compound	Color	Monosaccharide (yes/no)
water	_____	_____
glucose	_____	_____
fructose	_____	_____
sucrose	_____	_____
lactose	_____	_____
maltose	_____	_____
starch	_____	_____
glycogen	_____	_____

D. SELIWANOFF'S TEST FOR KETOHEXOSES

Compound	Color and Time	Ketohexose/Aldohexose
water	_____	_____
glucose	_____	_____
fructose	_____	_____
sucrose	_____	_____
lactose	_____	_____
maltose	_____	_____
starch	_____	_____
glycogen	_____	_____

E. IODINE TEST FOR POLYSACCHARIDES

Compound	Color	Polysaccharide(yes/no)
water	_____	_____
glucose	_____	_____
fructose	_____	_____
sucrose	_____	_____
lactose	_____	_____
maltose	_____	_____
starch	_____	_____
glycogen	_____	_____

F. HYDROLYSIS OF DI - AND POLYSACCHARIDES

Compound   Benedict's (F-1)   Iodine(F-2)

sucrose(hydrolyzed)   _____   _____

starch(hydrolyzed)   _____   _____

Compare the results of the Benedict's test with sucrose before and after hydrolysis.   Why are they different?

Compare the results of the Benedict's test with starch before and after hydrolysis. Why are they different?

Compare the results of the iodine test for starch before and after hydrolysis. Why is there a difference?

G. FERMENTATION TEST

Carbohydrate   Observations   Fermented(yes/no)

glucose   _____   _____

fructose   _____   _____

sucrose   _____   _____

lactose   _____   _____

maltose   _____   _____

starch   _____   _____

glycogen   _____   _____

What gas accumulated in the test tubes? _____

EXPERIMENT 37   CARBOHYDRATES   NAME_____

H. TESTING AN UNKNOWN CARBOHYDRATE

Test   Test Results   Conclusions

Benedict's   _____   _____

Barfoed's   _____   _____

Seliwanoff's   _____   _____

Iodine   _____   _____

Fermentation   _____   _____

Identification of Unknown _____

I. TESTING FOODS FOR CARBOHYDRATES

TEST RESULTS

FOOD	BENEDICT'S	BARFOED'S	SELIWANOFF'S	IODINE	FERMENTS
1._____	_____	_____	_____	_____	_____
2._____	_____	_____	_____	_____	_____
3._____	_____	_____	_____	_____	_____

POSSIBLE CARBOHYDRATE IN THE FOOD SAMPLES.

                              FOOD SAMPLE

              1.   2.   3.

Glucose   _____   _____   _____

Fructose   _____   _____   _____

Sucrose   _____   _____   _____

Lactose   _____   _____   _____

Maltose   _____   _____   _____

Starch   _____   _____   _____

Glycogen   _____   _____   _____

287

EXPERIMENT 38   LIPIDS

PURPOSE

1. Observe some physical and chemical properties of lipids.
2. To distinguish between saturated and unsaturated fats.
3. Observe saponification of triacylglycerol in the preparation of soap.
4. Test for reactions of soap with soft water, oil and $CaCl_2$.

MATERIALS

organic model kits
fatty acids: stearic acid, oleic acid, linoleic acid
lipids: olive oil, cottonseed oil, safflower oil, lard,
        Crisco(shortening), glycerol, coconut oil, soybean oil
        cholesterol, lecithin, vitamin A or cod liver oil
test tubes
5% Br_2 in CCl_4 solution
methylene chloride, CH_2Cl_2
fat or oil for soap: lard, Crisco or commerical shortening,
                      cottonseed oil, coconut oil, etc.
125-mL Erlenmeyer flask
boiling chips
hexane
95% ethanol
20% NaOH
aluminum foil
400-mL beaker
saturated NaCl solution
Buchner filtration apparatus
commercial detergent or soap
pH paper
5% $CaCl_2$
5% $MgCl_2$

KEYED OBJECTIVES IN TEXT: 16-1, 16-2, 16-3, 16-4, 16-5, 16-6

DISCUSSION OF EXPERIMENT

Lipids are a family of compounds that are grouped by similarities in solubility rather than structure. As a family, lipids are more soluble in nonpolar solvents such as ether, chloroform, or benzene. Most are not soluble in water. Important types of lipids include the triacylglcerols(fats and oils), phospholipids and the steroids. Compounds classified as lipids include fat-soluble vitamins A,D,E and K, cholesterol, hormones, portions of cell membranes and vegetable oils.

EXPERIMENT 38   LIPIDS

TRIACYLGLYCEROLS

The triacylglycerols are a subgroup of lipids consisting of glycerol that has formed ester bonds with three fatty acids. When the respective fatty acids (long-chain carboxylic acids found in fats) contain double bonds, the fat or oil is unsaturated.

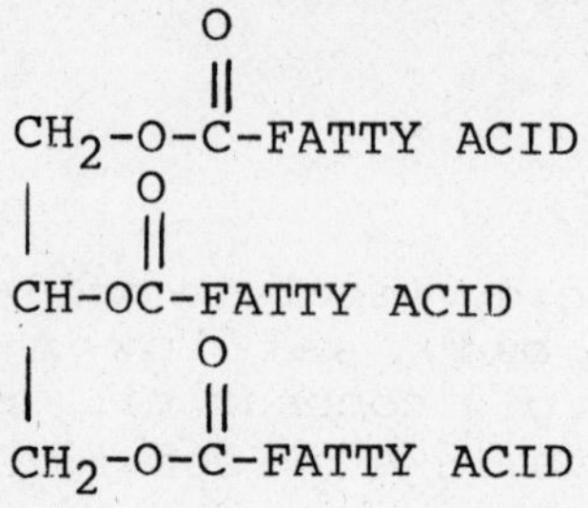

$$
\begin{array}{l}
\text{CH}_2\text{-O-C-FATTY ACID} \\
\text{CH-OC-FATTY ACID} \\
\text{CH}_2\text{-O-C-FATTY ACID}
\end{array}
$$

TRIACYLGLEROL

"TRIGLYCERIDE"

FATTY ACIDS

$$
\text{CH}_3(\text{CH}_2)_{16}\text{COH}
$$

STEARIC ACID
saturated

$$
\text{CH}_3(\text{CH}_2)_7\text{CH=CH(CH}_2)_7\text{COH}
$$

OLEIC ACID
unsaturated

Typically the saturated fats are solid at room temperature. Unsaturated fats containing some or many unsaturated fatty acids are usually found in the liquid form at room temperature.

BROMINE TEST FOR UNSATURATION

The presence of unsaturation in a fatty acid or a triacylglycerol can be detected by the bromine test performed in an earlier experiment for double bonds. If the orange color of the bromine solution fades quickly, an addition reaction has occurred and the oil or fat is unsaturated. The relative levels of unsaturation can be estimated by the amount of bromine solution added in order to achieve a permanent orange color with the fat or oil.

EXPERIMENT 38   LIPIDS

SAPONIFICATION: MAKING A SOAP

The reaction of a triacylglycerol with a base such as NaOH is called **saponification**. The fat or oil is hydrolyzed forming glycerol and the salts of the fatty acids which are soaps. The soap is precipitated out by the addition of a saturated NaCl solution. Several tests will be made with the soap you prepare such as measuring its pH, its ability to form suds in soft and hard water, and its reaction with oils.

$$
\begin{array}{l}
\text{CH}_2\text{OC-FATTY ACID} \\
\quad\;\; \text{O} \\
\text{CHOC-FATTY ACID} \\
\quad\;\; \text{O} \\
\text{CH}_2\text{OC-FATTY ACID}
\end{array}
\; + \; 3\,\text{NaOH} \;\longrightarrow\;
\begin{array}{l}
\text{CH}_2\text{OH} \\
\\
\text{CHOH} \\
\\
\text{CH}_2\text{OH}
\end{array}
\; + \; 3\,\text{Na}^+\,\text{FATTY ACID}^-
$$

TRIACYLGLYCEROL   GLYCEROL   SOAP (SODIUM SALTS
                                                            OF FATTY ACIDS)

LABORATORY ACTIVITIES

A. STRUCTURE OF TRIACYLGLYCEROLS

Using the organic model kits, make a model of glycerol and three propanoic acids. Combine the glycerol and propanoic acids by forming three ester bonds and removing three molecules of water. The resulting triacylglycerol is called tripropanoin or glycerol propanoate.

   The hydrolysis of the triacylglycerol you formed can be seen by breaking apart the ester bonds and adding the components of three water molecules.   Write an equation for the formation and hydrolysis of the triacylglycerol. Use double arrows to indicate that it is a reversible reaction.

B. SOLUBILITY OF SOME LIPIDS

Identify the following lipids as fatty acid, saturated or unsaturated fat, phospholipid, steroid or terpene.

      stearic acid, safflower oil, Crisco(or shortening)
      lecithin, cholesterol, vitamin A(puncture capsules or use
                                    cod liver oil)

291

Place 2 mL water in each of six test tubes. Add 5 drops of the liquid lipids, or the amount of solid lipid that you obtain on the tip of a microspatula. Stopper and shake each test tube. Record the solubility of each lipid. Repeat the solubility tests using 2 mL of hexane.

C.   TEST FOR UNSATURATION

CAUTION: AVOID CONTACT WITH THE BROMINE SOLUTION; IT CAN CAUSE PAINFUL BURNS.   WEAR GOGGLES AND WORK IN THE HOOD.

Place 5 drops of each lipid as listed in the laboratory record in separate test tubes. For solid lipids, use 0.1 g and dissolve in 1 mL methylene chloride. Add 2 mL of methylene chloride, CH_2Cl_2, to each sample.   Add the 5% bromine solution, drop by drop, until a permanent orange color remains. Count the number of drops of bromine added. Record.

D.   SAPONIFICATION: PREPARATION OF SOAP

CAUTION: OIL AND ETHANOL WILL BE HOT, AND MAY SPLATTER OR CATCH FIRE.   NaOH IS CAUSTIC AND CAN CUASE PERMANENT EYE DAMAGE. WEAR GOGGLES AT ALL TIMES.

1.   Place 6-7 g of a fat (lard, Crisco(shortening), cottonseed oil, olive oil, etc.) in a 125-mL Erlenmeyer flask. Add 20 mL 95% ethanol and 20 mL of 20% NaOH. CAUTION: 20% NaOH is corrosive. Use care in its use. Add a few boiling chips to the mixture to help prevent excessive foaming.

2.   Cover the mouth of the flask with aluminum foil. Poke a hole in the foil. Place the flask in a hot water bath using a 400-mL beaker approximately half-full of water. Heat for 30 minutes or until the solution becomes clear. There should be no separation of layers in the flask. If foaming is excessive, reduce the heat.

3.   Carefully place a few drops of the saponified solution in a test tube. Add 10 mL cold water. If fat droplets form, add 5 mL 20% NaOH and 5 mL ethanol to the flask and continue to heat carefully for 10 more minutes. When saponification is complete (solution is clear), let the mixture cool.

4.   Obtain 100 mL of saturated NaCl solution (30 g NaCl in 100 mL water). Pour the soap mixture into the salt solution. Soap should form and float at the surface. Collect the solid soap, using a filtration apparatus. Pour as much of the liquid as you can through the filter before you collect the solid soap precipitate. Wash the soap with two 15 mL portions of distilled water. Use plastic gloves to carefully pack the solid pieces of soap together to form a small cake. HANDLE CAREFULLY: The soap may still contain NaOH which can irritate the skin.

EXPERIMENT 38   LIPIDS

E.   REACTIONS OF SOAPS

Prepare a solution of your soap by dissolving 1 g in 50 mL of distilled water. Also prepare a solution of a commercial soap or detergent (Tide, All, Cheer, etc.) in the same amounts.

E-1   Place 10 mL of each solution in separate test tubes. Stopper and shake for 10 seconds. Layers of foam should form. Describe the results.

E-2   Using pH paper, determine the pH of each soap solutions in E-1. Record results.

E-3   Place 10 ml of each soap solution in separate test tubes. In a third test tube, place 10 mL of distilled water. Add 5 drops of safflower oil to each. Stopper and shake the test tubes. After 5 minutes, observe the contents of each test tube. Record the results.

E-4   Place 10 mL samples of each soap solution in separate test tubes. To each, add 2 mL of a 5% $CaCl_2$. Stopper and shake each solution. Record your observations.

       Repeat using 2 mL of 5% $MgCl_2$.

NAME_____

SECTION_____

DATE_____

LABORATORY RECORD

A. EQUATION FOR THE FORMATION AND HYDROLYSIS OF A TRIACYLGLYCEROL

B.   SOLUBILITY

LIPID	TYPE	SOLVENTS	
		WATER	HEXANE
Stearic acid	_____	_____	_____
Safflower oil	_____	_____	_____
Crisco	_____	_____	_____
Lecithin	_____	_____	_____
Cholesterol	_____	_____	_____
Vitamin A	_____	_____	_____

Explain why the above compounds are classified as lipids.

Why type of solvent is needed to remove an oil spot? Why?

C.   TEST FOR UNSATURATION

FATTY ACIDS:	Drops of Bromine Solution
stearic acid	_____
oleic acid	_____
linoleic acid	_____

TRIACYLGLYCEROLS (FATS AND OILS)	
cottonseed oil	_____
olive oil	_____
safflower oil	_____
Crisco (shortening)	_____

Write the structures of stearic acid, oleic acid, and linoleic acid.

Stearic acid

Oleic acid

Linoleic acid

Which is the most unsaturated fatty acid?   _____

Which is the most saturated fatty acid?   _____

Which is the most unsaturated triacylglycerol?_____

Which is the most saturated triacylglycerol?_____

. **SAPONIFICATION: FORMATION OF SOAP**

Write an equation for saponification of tristearin (glyceryl tristearate).

E.   REACTIONS OF SOAPS

Commercial product   _____

| TESTS | OBSERVATIONS | |
| | soap (prepared) | commercial product |

E-1 Shaking   _____   _____

E-2 pH   _____   _____

E-3 Oil   _____   _____

E-4 CaCl$_2$ _____   _____

     MgCl$_2$ _____   _____

What is the effect of a soap on a layer of oil?

How does soap react with the salts CaCl$_2$ or MgCl$_2$?

EXPERIMENT 39   AMINO ACIDS

PURPOSE

1. Identify the R group in amino acids.
2. Determine the pH of various amino acids in water.
3. Use chromatography to separate amino acids in a mixture.
4. Calculate R_f value for amino acids.
5. Use R_f values to identify amino acids.

MATERIALS

organic model kit
amino acids (1% solutions): alanine , glutamic acid, valine,
                            serine, aspartic acid, lysine,
                            tryptophan, unknown

Aspartame, artifical sweetener

small beaker (100 or 125 mL)
large beaker (400 or 600 mL)
plastic wrap
chromatography paper, Whatman #1 (12 cm x 20 cm)
solvent: n-butanol: glacial acetic acid: water (3:1:1)
drying oven (100°C)
2% ninhydrin spray
thin-layer chromatography plates

KEYED OBJECTIVES IN TEXT: 17-1, 17-2, 17-4, 17-5, 17-6, 17-7,17-8

DISCUSSION OF EXPERIMENT:

Amino acids and proteins are an important part of our total diet.
Proteins are broken down in the digestive process producing amino
acids needed to build tissues, enyzmes, skin and hair. All amino
acids are similar in structure. They contain an amino group and a
carboxylic acid group. Individual amino acids are differentiated
by variations in the R group, a group of atoms attached to the
central carbon atom. Variations in the R group account for
differences in the polarities and pH of the amino acids. Table 39-
1 lists the R group for some typical amino acids.

$$NH_2 - \overset{\overset{\displaystyle R}{|}}{\underset{\underset{\displaystyle H}{|}}{C}} - C\overset{\displaystyle O}{\underset{\displaystyle OH}{\diagup}}$$

amino group   carboxylic acid group

EXPERIMENT 39   AMINO ACIDS

TABLE 39-1   SOME TYPICAL AMINO ACIDS

R	Amino Acid	Abbreviation
H-	Glycine	Gly
CH_3-	Alanine	Ala
(see structure) $\underset{HO}{\overset{O}{\|}}C-CH_2-CH_2-$	Glutamic acid	Glu
(see structure) $\underset{HO}{\overset{O}{\|}}C-CH_2-$	Aspartic acid	Asp
$HO-CH_2-$	Serine	Ser
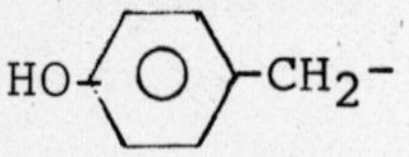 $HO-\langle\bigcirc\rangle-CH_2-$	Phenylalanine	Phe
$SH-CH_2-$	Cysteine	Cys
$\underset{CH_3}{\overset{CH_3}{>}}CH-$	Valine	Val
$\langle\bigcirc\rangle-CH_2-$	Tyrosine	Tyr
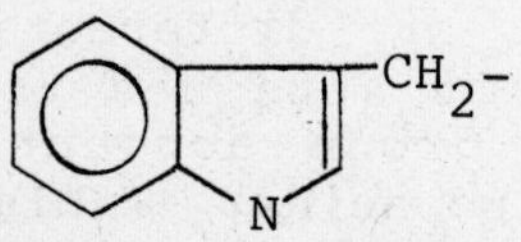 (indole) $-CH_2-$	Tryptophan	Trp
$NH_2-CH_2-CH_2-CH_2-CH_2-$	Lysine	Lys

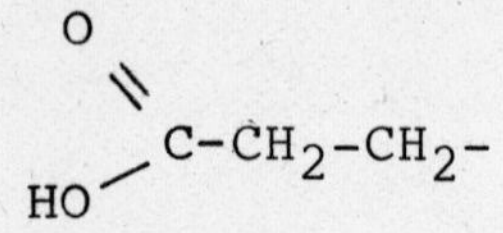

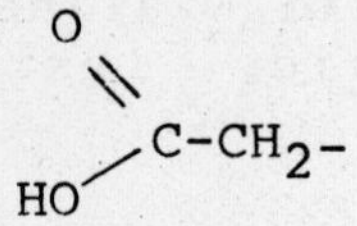

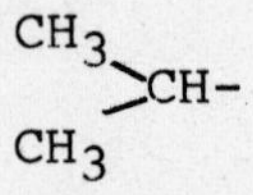

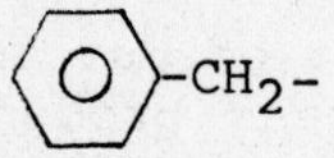

300

EXPERIMENT 39   AMINO ACIDS

DIPEPTIDES

A dipeptide is formed when two amino acids bond together by
forming a peptide bond. In a peptide bond, the amino group of one
amino acid bonds with the carboxylic acid group of the next amino
acid. In a protein, many amino acids bond together to form a
polypeptide. The specific order of amino acids determines the type
of protein that results.

$$
\underset{\text{Amino acid(1)}}{NH_2-\underset{\underset{H}{|}}{\overset{\overset{R_1}{|}}{C}}-\overset{\overset{O}{\parallel}}{C}-OH} \quad + \quad \underset{\text{Amino acid(2)}}{H-\underset{\underset{H}{|}}{\overset{\overset{R_2}{|}}{N}}-\underset{H}{\overset{O}{\parallel}}{C}-OH} \quad \longrightarrow \quad \underset{\text{Dipeptide}}{NH_2-\underset{\underset{H}{|}}{\overset{\overset{R_1}{|}}{C}}-\overset{\overset{O}{\parallel}}{C}-\underset{H}{\overset{}{N}}-\underset{H}{\overset{\overset{R_2}{|}}{C}}-\overset{O}{\parallel}{C}-OH}
$$

peptide bond

CHROMATOGRAPHY OF AMINO ACIDS

Different amino acids move at different rates in an organic
solvent as they move up a chromatography paper or a thin-layer
plate. Paper chromatogrphy uses filter paper while thin-layer
chromatography uses a piece of plastic or glass coated with a thin
layer of alumina or silica. Spots of the amino acids are placed
near the bottom of the paper or plate. Then the chromatogram is
placed in a solvent whose components travel up the paper or the
solid thin layer on the plate.

As the solvent travels, it carries the amino acids with it up
the chromatogram. Since the amino acids have different
solubilities in the solvent and in the stationary phase(paper),
they will separate from each other. Their location can be detected
by spraying a dried chromatogram with ninhydrin.

After visualization of the amino acids, the distance they
traveled from the starting line can be measured. The relationship
of the distance traveled by an amino acid compared to the distance
travled by the solvent is called the R_f value.

$$R_f = \frac{\text{distance traveled by amino acid}}{\text{distance traveled by solvent}}$$

When an unknown amino acid is to be identified, its R_f value
and color with ninhydrin can be matched to the R_f values of known
amino acids. In this way, the amino acids present in an unknown
mixture of amino acids or a hydrolysate of a dipeptide such as
Aspartame can be identified.

301

EXPERIMENT 39   AMINO ACIDS

LABORATORY ACTIVITIES

A.   STRUCTURE OF AMINO ACIDS AND DIPEPTIDES

Use the organic model kit to construct one of the amino acids from
Table 39-1. Draw its structure in the laboratory record. Give
its name and abbreviation.

Form a dipeptide by combining your amino acid with the amino
acid of one of your neighbors in lab. A molecule of water should
result when you construct the dipeptide. Draw the structure. Give
its name and abbreviation.

B.   pH OF AMINO ACIDS

Place 5-10 drops of the following 1% amino acid solutions on a
spot plate or on a piece of wax paper: alanine, lysine, glutamic
acid, serine, and aspartic acid. Use pH paper to determine the
pH of each amino acid solution. Draw the structure of each amino
acid and explain how its pH is related to its R group.

C.   PAPER CHROMATOGRAPHY

Preparation of Dipeptide Hydrolysate OPTIONAL Place 1 g Aspartame
in a 125-mL Erlenmeyer flask. Add 30 mL 6M HCl and place the flask
in a boiling water bath. Use a clamp to support the flask. Heat
at a gentle boil for 30 minutes.

Preparation of Chromatograph Tank   WORK IN THE HOOD: Prepare the
chromatography tank. Pour about 10 mL of the butanol:HAc:water
solvent into the large beaker to a depth of 1 cm. The depth of the
solvent must not exceed the origin line on your chromatogram.
Cover with plastic wrap.

Preparation of Paper Chromatogram   Obtain a piece of Whatman
No. 1 chromatography paper (12 cm x 20 cm). (The paper should
fit into the tank when it forms a cylinder.) Keep your fingers
off the paper as much as possible since you will transfer amino
acids from your skin. Handle the paper at the edge or with
plastic gloves. Draw a pencil line 2 cm from the long edge of
the paper. This is the starting or origin line. Mark seven
points, evenly spaced, along the line. (See Figure 39-1)

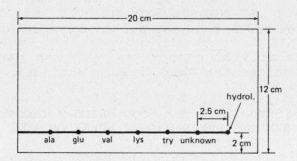

Figure 39-1   Preparation of a chromatogram.

Application of Amino Acids   Using the toothpick applicators or capillary tubes provided in each amino acid solution, lightly touch the tip to the paper. Keep the diameter of the spot as small as possible, like the size of the letter o.   Always return the applicator to the same amino acid solution. Label each spot as you go along, using a pencil. Apply a spot of the unknown amino acid and a spot of your dipeptide hydrolysate. A hair dryer can be used to dry the spots. When dry, repeat the application of amino acids, unknown, and protein hydrolysate two more times.

Running the Chromatogram   Form a cylinder with the paper. Join the edges using a stapler or masking tape without overlapping the paper. Place the paper cylinder in the chromatography tank, making sure that the cylinder does not touch the sides.   The level of the solvent must be lower than the origin line of amino acids.   See Figure 39-2.   Cover the beaker with the plastic wrap.   Do not disturb the tank as the solvent begins to flow up the chromatogram.

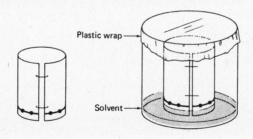

Figure 39-2   Chromatogram in solvent tank.

<u>Visualization</u> <u>of</u> <u>Amino</u> <u>Acids</u>   After about 1 hour, or when the
solvent is close to the top of the chromatogram, remove the paper
carefully.   DO NOT LET THE SOLVENT RUN OVER THE TOP OF THE PAPER.
Remove the staples and spread the paper out on a paper towel IN THE
HOOD WITH VENTILATION. <u>Mark</u> <u>the</u> <u>solvent</u> <u>line</u> <u>immediately</u> <u>using</u> <u>a</u>
<u>pencil.</u>   Let the solvent evaporate.   Dry the paper in a drying
oven (85°-100°C) or use the hair dryer.   When dry, spray the paper
evenly with ninhydrin.

CAUTION: SPRAY NINHYDRIN IN THE HOOD USING A LIGHT SPRAY.
DO NOT BREATHE FUMES OR GET SPRAY ON YOUR SKIN.

Allow the paper to dry. Place the paper in the drying oven for
3-5 minutes.   Colored spots will develop where the ninhydrin
reacted with the amino acids.   Outline each spot with a pencil.
Place a dot at the center of each spot. Record the color of each
spot.

<u>Calculation</u> <u>of</u> R$_f$ <u>Values</u>   The R$_f$ value represents the ratio of the
distance traveled by an amino acid compared to the distance
traveled by the solvent.   Measure the distance(mm) from the
starting line (origin) to the center dot of each spot to obtain the
distance traveled by each amino acid. Measure the distance from
the starting line to the solvent line to obtain the distance
traveled by the solvent.   See Figure 39-3.   Record in laboratory
record.   Calculate the R$_f$ values for the known amino acid samples,
the unknown amino acid, and the amino acids that appear in the
dipeptide hydrolysate.

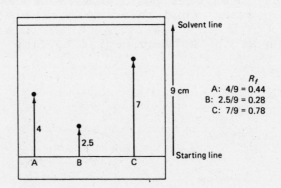

Figure 39-3   Developed chromatogram.   (R$_f$ values calculated for
         A,B, and C.)

Identification of Unknown Amino Acids   Compare the color and R_f values produced by the unknown amino acid and the hydrolysate. Identical amino acids will travel the same distance, give the same R_f values and form the same color with ninhydrin.   Identify the amino acids in the unknown and in the dipeptide.

D.   THIN-LAYER CHROMATOGRAPHY   OPTIONAL

Preparation of Thin-Layer Chromatogram   Obtain a precoated thin-layer chromatography sheet or plate. Handle only the top of the plastic or plate with your fingers. Spot the amino acids as you did for the paper chromatogram.

Preparation of Chromatography Tank   Obtain a container that will fit the thin-layer plastic sheet or glass plate. Place the solvent in the tank to a depth of 1 cm.

Running the Chromatogram   Set the thin-layer chromatogram in the tank with the top of the plate resting against the side of the tank. Cover the tank. Do not disturb the tank while solvent is traveling up the chromatogram. When the solvent nears the top of the plate, remove the plate. Mark the top of the solvent line, and let the plate dry in the hood.

Proceed with visualization of the amino acids and the calculations of the R_f values as you did in paper chromatography.

NAME_____

SECTION_____

LABORATORY RECORD

DATE_____

A. STRUCTURE OF AMINO ACIDS

Draw the structure of the amino acid in your model.

Name_____ Abbreviation_____

Draw the structure of the dipeptide in your combined model.

Name_____ Abbreviation_____

B. pH OF AMINO ACIDS

AMINO ACID	pH	STRUCTURE	EXPLANATION
Alanine			
Lysine			
Glutamic acid			
Serine			
Aspartic acid			

307

C.   PAPER CHROMATOGRAPHY

Attach or sketch the results of the paper chromatogram:

Calculations: R_f Values:

Distance from origin to final solvent line _____

AMINO ACID	COLOR	DISTANCE TRAVELED	R_f
Alanine	_____	_____	_____
Glutamic acid	_____	_____	_____
Valine	_____	_____	_____
Lysine	_____	_____	_____
Tryptophan	_____	_____	_____
Unknown	_____	_____	_____
	_____	_____	_____
Dipeptide	_____	_____	_____
	_____	_____	_____

Amino Acid(s) in Unknown_____

Amino Acids in Dipeptide Hydrolysate_____

D. THIN-LAYER CHROMATOGRAPHY

Sketch the results of the thin-layer chromatogram:

Calculations: R_f Values:

Distance from origin to final solvent line _____

AMINO ACID	COLOR	DISTANCE TRAVELED	R_f
Alanine	_____	_____	_____
Glutamic acid	_____	_____	_____
Valine	_____	_____	_____
Lysine	_____	_____	_____
Tryptophan	_____	_____	_____
Unknown	_____	_____	_____
	_____	_____	_____
Dipeptide	_____	_____	_____
	_____	_____	_____

Amino Acid(s) in Unknown_____

Amino Acids in Dipeptide Hydrolysate_____

EXPERIMENT 40   PROTEINS

PURPOSE

1. Identify the structural patterns of proteins.
2. Use the isoelectric point of casein in milk to
    isolate the protein.
3. Use chemical tests to identify amino acids and proteins.
4. Observe the denaturation of proteins.

MATERIALS

1% amino acids solutions: alanine, lysine, glutamic acid
                              serine, aspartic acid, glycine,
                              cysteine, tyrosine, and tryptophan

spot plates   0.5% $CuSO_4$
pH paper   10% NaOH
250-mL Erlenmeyer flask   0.1% ninhydrin solution
nonfat milk   conc. HNO_3
hot water bath   1% $PbAc_2$
thermometer   1% $AgNO_3$
10% acetic acid   Ethanol
beakers   Aspartame
Buchner filtration apparatus 95% ethanol

2% solutions of proteins: gelatin, albumin
casein (Part A)

DISCUSSION OF EXPERIMENT

Proteins are composed of about twenty amino acids, ten of which
must be obtained in the diet (essential amino acids). Proteins have
specific, three-dimensional structures determined by the order of
amino acids. The peptide bonds that bond the amino acids together
are the first or primary level of protein structure. Secondary
structure includes the -helix coiling of the peptide chain, or a
pleated sheet structure. Tertiary structures occurs when the side
groups interact(salt bridges, hydrophobic bonds, disulfide bonds,
etc.) to pull the chain into a specific, three-dimensional shape.
The quaternary structure consists of an active protein that has two
or more even number of tertiary units in a group.

TESTS FOR AMINO ACIDS AND PROTEINS

There are several reactions of amino acids and/or proteins that give specific colors that can be used to test for the presence of amino acids or proteins.

Biuret Test

The biuret test is positive for proteins with two or more peptide bonds. When a solution of cupric sulfate and NaOH is added to a protein of three or more amino acids, a pink to violet color will appear. Individual amino acids will not give the biuret test, and the solution will remain blue.

Ninhydrin Test

The ninhydrin test produces a blue-violet color with amino acids and proteins. Proline and hydroxyproline give a yellow color.

Xanthoproteic Test

The xanthoproteic test is specific for the R groups of tyrosine and tryptophan. Amino acids or proteins containing aromatic rings will react with concentrated nitric acid to give nitro-substituted benzene rings which appear as yellow-colored products.

Sulfur Test

A positive sulfur test occurs in a basic solution when the sulfur in the side chains of cysteine or cystine group is converted to a sulfide and forms black PbS precipitate when lead acetate is added.

DENATURATION OF PROTEINS

Denaturation of a protein occurs when the tertiary or quaternary structure of the protein is altered or destroyed. In many cases, coagulation of the protein occurs. Agents causing denaturation include heat, acid, base, ethanol, tannic acid, and heavy metal ions of silver, lead, and mercury. Heat increases the vibrations of the atoms which disrupts the hydrogen bonding and hydrophobic(nonpolar-nonpolar) bonds. Strong acids and bases add or remove hydrogen ions to the ionic side groups. As a result, salt bridges (ionic bonds) are broken. Alcohol is an organic solvent that disrupts the hydrogen bonding of the secondary and tertiary protein structures. The heavy metal ions, Ag^+, Pb^{2+}, and Hg^{2+}, react with the disulfide bonds of the tertiary structures and the carboxylic groups of the acidic R groups.

312

EXPERIMENT 40   PROTEINS

LABORATORY ACTIVITIES

A.   SEPARATION OF A PROTEIN (CASEIN) FROM MILK

A typical source of protein is milk, which contains the protein casein. When nonfat milk is acidified and the isoelectric point is reached, the protein separates out of the solution. The amount of protein will be determined and the percentage of casein in milk calculated.

A-1 Weigh a 250-mL Erlenmeyer flask. Record the mass. Add 50 mL of nonfat milk to the flask and reweigh.

A-2 Place the flask in a hot water bath made from a 400-mL beaker. See Figure 40-1. Hold the temperature of the milk at about 50°C, while you slowly add 10% acetic acid dropwise. When the casein reaches its isoelectric point, it becomes insoluble, and precipitates out leaving a clear solution. When no further precipitation occurs, stop adding acid. Let the sample cool.

Use the filtration apparatus to collect the protein particles. Let the protein dry on a paper towel or piece of filter paper. Weigh and record the mass of protein from the milk sample.

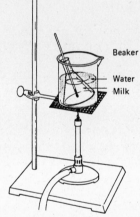

Figure 40-1 Apparatus for heating milk to separate casein.

A-3 Calculate the percentage of casein in nonfat milk.

% casein   =   mass casein   x 100%
                mass of milk

B. BIURET TEST FOR PROTEINS

Place 2 mL or a small amount of crystals on the tip of a spatula of the following in five separate test tubes:

> glycine, alanine, gelatin, Aspartame, egg albumin and casein obtained in Part A. (Aspartame is an artificial sweetner that is a dipeptide.)

Add 2 mL of 10% NaOH and mix thoroughly. Add 5 drops 0.5% cupric sulfate and mix. Record the appearance of each solution. The formation of a pink-violet color indicates the presence of a protein with two or more peptide bonds. If such a protein is not present, the blue color of the cupric sulfate will remain. Record the results and your conclusions.

C.   NINHYDRIN TEST

Place 3 mL or a small amount of solid of the following in separate test tubes: glycine, alanine, tyrosine, gelatin, albumin, Aspartame, and casein from Part A. Add 1 mL of ninhydrin solution to each sample. Place in a boiling water bath for 5 minutes. Record the results and your conclusions.

D.   XANTHOPROTEIC TEST

Place 2 mL of the following solutions or a small amount of solid on the tip of a spatula in separate test tubes:

> tryptophan, tyrosine, gelatin, Aspartame,
> albumin, and casein from Part A.

Add 10 drops of concentrated HNO_3 to each sample. CAUTION: CONCENTRATED HNO_3 IS DAMAGING TO SKIN AND EYES. WEAR GOGGLES. USE CARE IN ITS USE. Record any color changes. Place the test tubes in cold water and let them cool. CAREFULLY add dropwise 10% NaOH until the solution is basic to red litmus. Record the results and your conclusions.

E.   SULFUR TEST

Place 2 mL or a small amount of solid of the following in separate tests tubes:alanine, cysteine, gelatin, albumin,
          and casein from Part C in separate test tubes.

Add 4 mL of 10% NaOH to each sample. Place the test tubes in a boiling water bath for 10 min. Add 5 drops of $PbAc_2$ solution. Record the results and your conclusions.

F.   DENATURATION OF PROTEINS

Prepare 5 test tubes each with 3 mL egg albumin solution. Use one sample for each test. Record your observations and give a brief explanation for the results.

Denaturation Agent	Procedure
Heat	Heat gently for 2-3 minutes.
Heavy Metals	(1) Slowly add 10 drops of $AgNO_3$. (2) Slowly add 10 drops of $PbAc_2$.
Strong Acid	Add 2 mL 6M HNO_3.
Alcohol	Add 10 mL 95% ethanol.

SECTION _____
LABORATORY RECORD   DATE_____

A.   SEPARATION OF A PROTEIN(CASEIN) FROM MILK

A-1   Mass of flask   _____

       Mass of flask and milk   _____

       Mass of milk   _____

A-2   Mass of casein product   _____

A-3   % casein in milk   _____

TESTS FOR AMINO ACIDS AND PROTEINS

B.   BIURET TEST

COMPOUND	COLOR	CONCLUSIONS
Glycine	_____	_____
Alanine	_____	_____
Aspartame	_____	_____
Gelatin	_____	_____
Albumin	_____	_____
Casein	_____	_____

C. NINHYDRIN TEST

COMPOUND	COLOR	CONCLUSIONS
Gycine	_____	_____
Alanine	_____	_____
Tyrosine	_____	_____
Aspartame	_____	_____
Gelatin	_____	_____
Albumin	_____	_____
Casein	_____	_____

EXPERIMENT 40   PROTEINS   NAME_____

D. XANTHOPROTEIC TEST

COMPOUND   COLOR   CONCLUSIONS

Tryptophan   _____   _____

Tyrosine   _____   _____

Gelatin   _____   _____

Aspartame   _____   _____

Albumin   _____   _____

Casein   _____   _____

E. SULFUR TEST

COMPOUND   COLOR   CONCLUSIONS

Alanine   _____   _____

Cysteine   _____   _____

Gelatin   _____   _____

Albumin   _____   _____

Casein   _____   _____

F. DENATURATION OF PROTEINS

                           OBSERVATIONS
DENATURING AGENT

Heat   _____

Heavy metals-AgNO$_3$_____

              PbAc$_2$_____

Strong acid   _____

Alcohol   _____

EXPERIMENT 40 PROTEINS NAME____________________

QUESTIONS AND PROBLEMS

1. Write the primary structure for the polypeptide of Ala-Gly-Cys-Phe-Gly.

2. Will the polypeptide in question 1 give a positive or negative result in the following tests for a protein or amino acid? Explain why or why not.

   a. Biuret

   b. Xanthoproteic

   c. Sulfur

   d. Ninhydrin

3. Why does a protein undergo denaturation?

4. Under what circumstances would the denturation of a protein be useful?

EXPERIMENT 41   ENZYME ACTIVITY

PURPOSE

1. Associate the presence of enzymes with the catalysis of chemical reactions in living cells.
2. Determine the effect of enzyme concentration, subtrate concentration, temperature, pH and heavy-metal salts upon the activity of salivary amylase.

MATERIALS

test tubes   1% starch (buffered pH 7.0)
test tube rack   iodine reagent
thermometer   0.1M $AgNO_3$
beakers
eyedropper   0.1M NaCl
spot plate, or wax paper   ethanol
temperature baths: crushed ice
warm water bath ($60^{O}C$)   buffers(pH 3,5,7,9,11)
boiling water bath($100^{O}C$)   iodine reagent
ice-water bath($0^{O}C$)

KEYED OBJECTIVES IN TEXT: 18-1, 18-2, 18-3, 18-4, 18-5, 18-6

DISCUSSION OF EXPERIMENT

Enzmyes are biological cataysts. They participate in essentially every biological reaction necessary for the maintenance of a living system. In this experiment, you will use a readily available enzyme, salivary amylase, which begins the hydrolysis of carbohydrates (amylose) in the saliva of the mouth.

An enzymes act upon the reactants or <u>substrates</u> of a reaction to give an <u>enzyme-substrate</u> intermediate. New compounds called <u>products</u> result. The enzymes are used over and over during a reaction.

Enzyme + Substrate   Enzyme-substrate   Enzyme + Product

   E   +   S $\rightleftharpoons$ ES $\longrightarrow$ E   +   P

Enzymes catalyze reactions in a cell at body temperature and mild conditions. The rates of enzyme-catalyzed reactions are much faster than they would be without the enzymes. Enzymes speed up a reaction by lowering the energy of activation required to make the reaction occur.

To follow the action of an enzyme, it is necessary to test for the appearance of a product, or the disappearance of a reactant over a measured period of time.

EXPERIMENT 41   ENZYME ACTIVITY

AMYLASE

Amylase is an enzyme that is found in the saliva. In the reactions of amylase with starch, you will test for the disappearance of starch by reacting samples of the reaction mixtures with iodine. Initially, a starch solution gives a blue-black color with iodine. The amylase catalyzes the hydrolysis of the α-1,4 glycosidic bonds forming smaller polysaccharides, dextrins, maltose and eventually glucose.

$$\text{Starch} \longrightarrow \begin{array}{l} \text{Smaller polysaccharides} \\ \text{Dextrins} \\ \text{Maltose} \end{array} \longrightarrow \text{glucose}$$

After the starch is hydrolyzed, the blue-black color produced with iodine no longer occurs, and only the red or gold color of the iodine solution is seen. The faster the amylase hydrolyzes the starch, the more quickly the blue-black color disappears. If the blue-black color does not fade, you may conclude that the enzyme is no longer active and that no hydrolysis of starch has occurred.

FACTORS AFFECTING ENZYME ACTIVITY

Enzyme activity depends upon several factors including enzyme concentration, substrate concentration, pH, temperature and heavy metals. Your own saliva containing the amylase enzyme will be used for this experiment although the levels of amylase vary considerably from one person to another. Each experiment must be timed. As you proceed with each experiment, you will check enzyme activity by reacting a few drops of the reaction mixture with iodine. The time at which the blue-black color of starch disappears will be noted in each experiment.

The time required for the disapparance of starch will be correlated to the relative enzyme activity. When enzyme activity is high, the time for the starch to disappear will be very short. When the enzyme is operating poorly or not at all, the activity will be low and more time will be required for the starch to disappear. In some cases, the enzyme will be completely inactivated and the blue-black color of starch and iodine will persist throughout the entire experiment. Graphs will be prepared showing the effects of concentration, pH, temperature and heavy metals upon the relative enzyme activity.

EXPERIMENT 41   ENZYME ACTIVITY

LABORATORY ACTIVITIES

A.   ENZYME CONCENTRATION

A-1 Collect 2-3 mL of saliva in a small, clean beaker. Add an
equal volume of distilled water to the saliva and mix.
Chewing gum or rubber bands may help your secretion of saliva.

Using 5 test tubes, place 5-mL portions of 1% starch
solution in each. Set up a water bath and heat to about 37°C.
If it get too warm, add tap water to achieve a water bath
temperature close to 37°C. Place the test tubes in the 37°C
water bath for 5 min.

Add the following amounts of saliva to the test tubes as
quickly as you can, mix thoroughly, and replace the test tubes
in the 37°C water bath. Record the time that you place the
test tubes in the 37°C water bath. Watch the temperature of
the water bath adding more warm water as needed.

Test Tube	Drops of Saliva Solution
1	1
2	5
3	10
4	15
5	30

Starch-Iodine Test

Prepare a spot plate or sheet of wax paper for testing for
starch in the samples. Place 1 drop of iodine reagent in each
depresssion of the spot plate or on the wax paper. Add a drop
of starch solution to the first drop of iodine. The reaction
should give a deep blue-black color which indicates the
presence of starch.

A-2 Five minutes after addition of enzyme, remove a drop of each
mixture (use clean droppers each time) and add to a drop of
iodine in the spot plate or on the wax paper. A blue-black
color indicates that starch is present after 5 min. If the
color with iodine is red or gold, the starch has been
completely hydrolyzed. Record the time for disappearance of
the blue-black color in any of the samples. Clean the spot
plate or wax paper.

Repeat the testing with fresh drops of iodine in the spot
plate or on the wax paper every 5 min thereafter. Continue
testing for 30 min. Record the time when the blue-black color
no longer appears for each sample.

323

A-3   Calculate the amount of time in minutes for the hydrolysis of starch in each sample.

A-4   Plot the time (minutes) for the starch to hydrolyze (disappearance of blue-black color) against the amount (drops) of saliva solution.

B. EFFECT OF pH

B-1   Place 5 mL of buffer solutions (pH 3,5,7,9,and 11) in separate test tubes. Label. Add 5 drops of saliva solution to each test tube. Place the test tubes all together in a 37^OC water bath for five min. Prepare a spot plate or wax paper with drops (1 each) of iodine for testing the reaction progress. To start the reaction, add 5 ml of 1% starch solution to each test tube. Record the time you added the starch solution.

Five minutes after the starch was added, remove a drop of the reaction mixture from each test tube and perform the starch-iodine test. Repeat the test for starch every 5 minutes thereafter. Continue testing as long as starch is present up to 30 minutes. In some test tubes, the enzyme will be inactive, and the test will be positive for starch the entire time.

B-2   Calculate the time required for the disappearance of starch (blue-black color) at each pH.

B-3   Plot the time for starch to disappear (no longer gives a blue-black color) against the pH of the samples.

C.   EFFECT OF TEMPERATURE

C-1   Work with your lab neighbors to prepare temperature baths using larger beakers as follows:

     0^OC   ice-water mixture
     20-25°C   tap water
     37^OC   warmed tap water
     70^OC   warm water bath
     100°C   boiling hot water bath

Place 5 mL of 1% starch solution in each of five test tubes. Place one test tube in each of the five water baths and leave for 5 min.   Then add 10 drops of saliva solution to each test tube, and record the time.   Leave the test tubes in the water baths another 5 minutes.

Five minutes after the first addition of enzyme, remove a drop from each test tube and test for starch . Test for starch every 5 min thereafter. Record the time at which the blue-black color disappears for each sample. Continue testing for 30 min.

C-2   Calculate the time (minutes) required for the disappearance of starch in each sample.

C-3   Plot the relative enzyme activity (hydrolysis time of starch) against the temperature.

D.   INHIBITION OF ENZYME ACTIVITY

D-1   Prepare 4 test tubes with the following:

Test tube

1	2 mL 0.1M $AgNO_3$
2	2ml 0.1M NaCl
3	2 mL ethanol
4	2 mL water

Add 10 drops of the salive solution to each. Place the test tubes in a 37°C water bath and leave for 5 min. Prepare a spot plate or wax paper with drops of iodine reagent. To start the reaction, add 4 mL of starch solution to each test tube and mix thoroughly.

D-2   Five minutes after the addition of enzyme, remove a drop of solution from each test tube and test for starch. Repeat the test every 5 min thereafter. Continue testing for 20 minutes.

D-3   Calculate the time required for the blue color to disappear.

EXPERIMENT 41　　　ENZYME ACTIVITY　　　NAME_____
　　　　　　　　　　　　　　　　　　　　　SECTION_____
LABORATORY RECORD　　　　　　　　　DATE_____

A.　ENZYME CONCENTRATION

TEST TUBE	DROPS OF ENZYME	A-1 STARTING TIME	A-2 TIME BLUE COLOR FADES	A-3 TIME FOR STARCH HYDROLYSIS
1	1	_____	_____	_____
2	5	_____	_____	_____
3	10	_____	_____	_____
4	15	_____	_____	_____
5	30	_____	_____	_____

A-4　GRAPH:　ENZYME ACTIVITY (TIME FOR STARCH HYDROLYSIS)
　　　　　　　VS. ENZYME CONCENTRATION

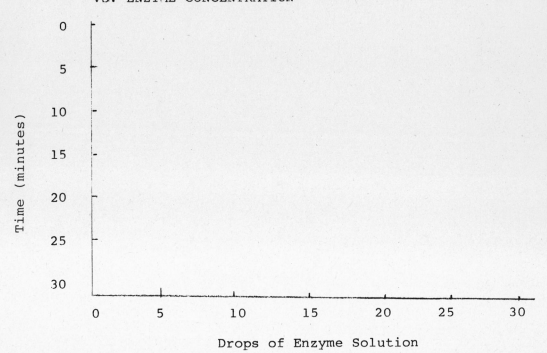

EXPERIMENT 41   ENZYME ACTIVITY   NAME_____

B. EFFECT OF pH

pH	B-1 STARTING TIME	TIME BLUE COLOR FADES	B-2 TIME FOR STARCH HYDROLYSIS
3	_____	_____	_____
5	_____	_____	_____
7	_____	_____	_____
9	_____	_____	_____
11	_____	_____	_____

B-3 GRAPH: ENZYME ACTIVITY(TIME FOR STARCH HYDROLYSIS) VS. pH

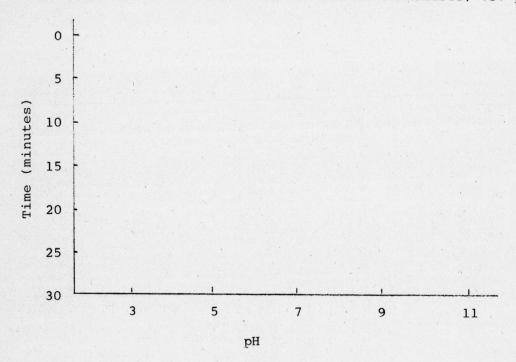

C.   EFFECT OF TEMPERATURE

TEMPERATURE	C-1 STARTING TIME	TIME FOR BLUE COLOR TO FADE	C-2 TIME FOR STARCH STARCH HYDROLYSIS
0°C	_____	_____	_____
15°C	_____	_____	_____
37°C	_____	_____	_____
60°C	_____	_____	_____
100°C	_____	_____	_____

C-3   GRAPH: ENZYME ACTIVITY(TIME FOR STARCH HYDROLYSIS) VS TEMP.

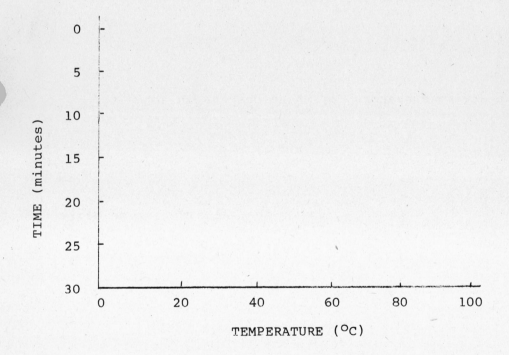

D.   INHIBITION OF ENZYME ACTIVITY

COMPOUND	D-1 STARTING TIME	D-2 TIME FOR BLUE COLOR TO FADE	D-3 TIME FOR STARCH HYDROLYSIS
$AgNO_3$	_____	_____	_____
NaCl	_____	_____	_____
Ethanol	_____	_____	_____
Water	_____	_____	_____

QUESTIONS

Refer to your results and graphs in parts A,C,D and D.

1. (A) How does the activity of the enzyme change as its
       concentration is increased?

   What are some reasons for these changes in activity?

2. (B) What is the effect of pH upon the relative enzyme activity?

   What are some reasons for this effect?

QUESTIONS (CONT)

3.(C) How is the enzyme affected by low temperatures? By high
      temperatures?

      What are some reasons for these temperature effects?

      According to your experiment, what is the optimum temperature
      for salivary amylase?

4.(D) Which of the compounds added to the reaction tubes are
      inhibitors?

      Why do these compounds act as inhibitors of enzyme activity?

      Why is an alcohol swab used prior to giving an injection to a
      patient?

EXPERIMENT 42   VITAMINS

PURPOSE

1. Identify a vitamin as water or fat-soluble.
2. Determine the vitamin C content in a variety of citrus juices and other solutions.
3. Determine the effect of heat upon vitamin C.

MATERIALS

Vitamins A,B,C,D,E (or as many as possible)
methylene chloride (CH_2Cl_2)
Vitamic C tablet
50-mL buret
6M HAc(Acetic Acid)
Fruit juices: orange, grapefruit, powdered, etc.
1% starch indicator
iodine reagent (16 g KI + 1.5 g I_2 in 1 L of solution)
Erlenmeyer flasks (250-mL)

KEYED OBJECTIVES IN TEXT: 18-5

DISCUSSION OF EXPERIMENT

Vitamins are organic compounds that are not synthesized by the cells of the body and must be provided in our diet. They often act as cofactors for certain enzymes. The deficiency of a vitamin in a diet can affect the activity of an enzyme bringing about a deficiency disease. A lack of vitamin C can cause scurvy, while a deficiency in vitamin A is associated with night blindness. A diet low in vitamin D can cause rickets in children, and a deficiency in B 12 can lead to anemia. Vitamins B and C which are soluble in water are called "water-soluble vitamins", while vitamins A, D, E, and K are "fat soluble vitamins".

In 1979, the National Academy of Sciences recommended the following daily dietary allowances for females 23-50 of age,120 lb.

Vitamin A	800 µg		Riboflavin	1.2 mg
Vitamin D	5 µg		Niacin	13 mg
Vitamin C	60 mg		Vitamin B_6	2.0 mg
Thiamin	1 mg		Vitamin B_{12}	3.0 µg
Folic acid	300 µg			

EXPERIMENT 42   VITAMINS

TESTING FOR VITAMIN C

    Vitamin C is used by the body to fight infection and repair
damaged tissues. It is present in a variety of foods including
citrus fruits.   Since vitamin C is a reducing agent, we can
measure its presence by its reaction with iodine.   The indicator
for this reaction is starch. When iodine is present, the starch
turns a deep-blue color. However, if vitamin C is present and it
has reduced the iodine to iodide, the starch does not react and no
deep-blue color forms.

$$I_2 \xrightarrow{\text{Vitamin C}} 2\ I^-$$

        blue with starch   no color with starch

LABORATORY ACTIVITIES

A. VITAMIN SOLUBILITIES

Place a small amount of each available vitamin in separate test
tubes. Add 2 mL of water to each sample. Mix and observe each
solution. If two layers form, the vitamin is not water soluble.
If the vitamin dissolves to give a clear solution, it is water
soluble. Record your observations. Repeat the procedure adding
2 mL methylene chloride (CH_2Cl_2) in place of water to each
vitamin sample. Record your observations.

B. STANDARDIZATION OF VITAMIN C

B-1 Weigh a 250-mL Erlenmeyer flask and record.

B-2 Crush a tablet of vitamin C (100 mg tablet) and place in the
     flask. Record the mass of the flask and the vitamin C.

B-3 Add 25 mL distilled water and mix.   Add 2 mL 6M HAc and 3 mL
     starch indicator. Set up a buret for titration. Fill the
     buret with prepared iodine solution. CAUTION: KEEP IODINE
     REAGENT AWAY FROM CLOTHES AND SKIN. Record the initial
     reading of the iodine solution in the buret.

B-4 Add the iodine to the flask containing the vitamin C tablet
     until a deep-blue color persists for 30 sec. This is the
     endpoint. Record the final reading of the iodine solution in
     the buret.

B-5 Calculate the mg vitamin C contained in the Erlenmeyer flask.

B-6 Calculate the volume of iodine solution.

B-7 Calculate the mass(mg) vitamin C that reacts with 1 mL iodine
     solution.

334

$$\frac{mg\ vitamin\ C}{mL\ iodine\ solution} = mg\ vitamin\ C/\ 1\ mL\ iodine\ solution$$

C. DETERMINATION OF VITAMIN C IN FRUIT JUICES

C-1 Place a 20 mL sample of a fruit juice in a clean Erlenmeyer flask. You may also use 1.0 g of a powdered fruit drink such as Tang. Record the type of fruit juice used. Add 50 mL of water and 2 mL 6M HAc. Add 3 mL 1% starch indicator.

C-2 Add iodine solution from your buret to the flask containing the fruit juice and starch indicator until a deep-blue color is obtained. Deeply colored juices may obscure this color. Record the volume of iodine solution added to reach the endpoint of the reaction.

C-3 Calculate the mg vitamin C in the sample. Use the value of mg vitamin C/ mL iodine solution obtained in step B-7.

$$mL\ iodine\ solution\ x\ \frac{mg\ vitamin\ C}{1\ mL\ iodine\ solution} = mg\ vitamin\ C\ in\ sample$$

C-4 Repeat the titration again and average your results, or try other fruit juices available in the lab. Record the results.

D. HEAT DESTRUCTION OF VITAMIN C

D-1 Obtain 20-mL of the juice that gave you the greatest amount of vitamin C content. Place the sample in a 250-mL Erlenmeyer flask. Add 50 mL of water. Using a low flame, gently boil the solution for 10 min. Remove the flask and cool it in an ice-water bath. Add 2 mL HAc and 3 mL starch indicator. Titrate with iodine solution to the deep-blue endpoint.

D-2 Calculate the mg Vitamin C present in the heated sample.

D-3 Calculate the mg Vitamin C lost(destroyed) by heating the sample.

D-4 If Vitamin C is still present, heat new 20-mL samples of the juice; one for 20 min and the other for 30 min. Cool the samples and titrate with iodine and starch indicator.

E. VITAMIN C IN URINE   (OPTIONAL)

Carry out the titration with iodine and starch with 25-mL samples of urine. Compare the vitamin C content in the urine of a student who took a vitamin C tablet that morning, one who drank a citrus juice, and one who took no source of vitamin C. You can also monitor the output of vitamin C over the day if laboratory time is available. You will be responsible for setting up your laboratory record and showing calculations.

EXPERIMENT 42   VITAMINS   NAME_____
                                         SECTION_____
LABORATORY RECORD   DATE_____

A. VITAMIN SOLUBILITIES

NAME OF VITAMIN	SOLUBILITY IN WATER	SOLUBILITY IN METHYLENE CHLORIDE	TYPE OF VITAMIN (WATER/FAT SOLUBLE)
_____	_____	_____	_____
_____	_____	_____	_____
_____	_____	_____	_____
_____	_____	_____	_____
_____	_____	_____	_____
_____	_____	_____	_____
_____	_____	_____	_____

B. STANDARDIZATION OF VITAMIN C

DATA:

B-1   Mass of flask   _____

B-2   Mass of flask and Vitamin C   _____

B-3   Initial buret reading   _____

B-4   Final buret reading   _____

CALCULATIONS:

B-5   Mass(mg) Vitamin C   _____

B-6   Volume of iodine solution   _____

B-7   mg Vitamin C/1 mL iodine solution_____
       Calculations:

C. DETERMINATION OF VITAMIN C IN FRUIT JUICE

SAMPLE	BURET READING INITIAL	BURET READING FINAL	VOLUME OF IODINE (mL)	mg VITAMIN C
1. _____	_____	_____	_____	_____
2. _____	_____	_____	_____	_____
3. _____	_____	_____	_____	_____
4. _____	_____	_____	_____	_____

Calculations of mg Vitamin C in each sample:

Sample _____

1. _____

2. _____

3. _____

4. _____

If the daily requirement is 60 mg vitamin C, how many milliliters (or grams) of each sample do you need to need the minimum daily requirement?

Sample _____

1. _____

2. _____

3. _____

4. _____

D. HEAT DESTRUCTION OF VITAMIN C

Sample _____

			(D-1)	(D-2)	(D-3)
HEATING TIME (min)	BURET READING Initial	Final	VOLUME IODINE	mg VITAMIN C	mg Vitamin C LOST
0					
10					
20					
30					

How does heat affect the vitamin C content of a fruit juice?

E. VITAMIN C IN URINE
DATA:

CALCULATIONS:

EXPERIMENT 43   DIGESTION OF FOODSTUFFS

PURPOSE

1. Identify digestion as a process of hydrolysis.
2. Determine the hydrolysis products of carbohydrate, fat and protein digestion.

MATERIALS

test tubes	50-mL buret
1% starch	0.1N NaOH
iodine solution	2% pancreatin
spot plate or wax paper	pH meter or pH paper
Benedict's reagent	phenolphthalein
safflower oil	hard-cooked egg
bile salts	2% pepsin
whole milk	0.1N HCl

KEYED OBJECTIVES IN TEXT: 18-6

DISCUSSION OF EXPERIMENT

The digestive processes utilize enzymes to carry out the hydrolysis of large molecules in our food to molecules small enough to dialyze through the intestinal wall into the blood or lymph.

STARCH DIGESTION

Starch makes up one of the major carbohydrates in our foods. In order to use starch, it must be hydrolyzed into glucose molecules. Digestion of starch begins in the mouth by the action of an enzyme, salivary amylase. Hydrolysis continues in the small intestine through the action of pancreatic amylase, maltase, sucrase and lactase.

$$\text{starch(amylose)} \xrightarrow{\text{carbohydrases (amylase, maltase)}} \text{glucose molecules}$$

LIPID DIGESTION

Approximately 30-40% of our diet consists of fats and oil (triacylglycerols). Chemically, a fat is an ester of glycerol and fatty acids. Digestion of fats begins in the intestine with bile salts and the enzymic action of lipases obtained from the gall bladder. The bile salts cause the fat to break up into smaller droplets (emulsification) increasing the surface area and the lipases hydrolyze the ester bonds of the fats.

$$\text{fats} \xrightarrow{\text{lipases}} \text{glycerol + fatty acids}$$

EXPERIMENT 43   DIGESTION OF FOODSTUFFS

PROTEIN DIGESTION

Proteins begin digestion in the stomach where HCl activates the proteases such as pepsin that hydrolyze peptide bonds. The resulting products of protein digestion are the amino acids.

                          proteases (pepsin, trypsin)
          proteins   amino acids

LABORATORY ACTIVITIES

A.   DIGESTION OF CARBOHYDRATES

Hydrolysis of Starch

Collect 1 mL of saliva in a small beaker. Prepare two test tubes each with 5 mL 1% starch. To one of the test tubes, add 10 drops of saliva containing salivary enzyme. Mix thoroughly. Prepare a spot plate or wax paper with 1 drop of iodine reagent for each test.

Every two minutes, place a drop of each mixture in the spot plate or on a drop of iodine on the wax paper. A deep-blue color indicates that starch is present. Continue testing the mixtures until the deep-blue color for starch no longer forms. Record the time.

A-2   Formation of Glucose

To test for the presence of glucose as a final product of starch digestion, add 5 mL of Benedict's reagent to each of the two test tubes and contents. Place the test tubes in a boiling water bath for 5 min. Record the colors that form. Determine if glucose is present or not.

B.   DIGESTION OF FATS

B-1   Bile Salts

Place 2 mL safflower oil in each of two test tubes. Add 8 mL water to one test tube. To the other sample, add 5 mL water and 3 mL bile salts. Mix thoroughly. Let the test tubes stand for 30 min. Record your observations.

B-2   Hydrolysis by Lipase

Place 50 mL of milk in an Erlenmeyer flask. Place the flask in a 37°C water bath using a 400 mL beaker. Set up a buret containing 0.1 M NaOH. Add 10 mL 2% pancreatin to the flask and mix thorougly.

Carefully pour out 10 mL of the milk mixture into another flask. Return the rest of the milk to the 37°C water bath. Test the pH of the milk sample with a pH meter or pH paper. Record pH.

Add 3-5 drops of phenolphthalein to the milk sample. Titrate with the 0.1 N NaOH from the buret until a permanent light pink color is obtained. This marks the endpoint. Record the number of milliliters of NaOH required to reach the endpoint.

Remove 10-mL samples of the milk and lipase mixture every 20 minutes until 60 min have elapsed. Immediately determine the pH of each sample and then titrate with 0.1N NaOH. Record the number of milliliters of NaOH used each time.

C. PROTEIN DIGESTION

Obtain three small pieces of the white part of a hard-boiled egg. Place the pieces in three separate test tubes. Prepare the three test tubes as follows:

test tube	Solutions
1	5 mL water + 1 mL 0.1N HCl
2	5 mL pepsin + 1 mL 0.1N HCl
3	5 mL pepsin + 1 mL water

Place the test tubes in a 37°C water bath for 1 hr. Record any changes in the egg white portion in each test tube.

A.   DIGESTION OF CARBOHYDRATES

A-1   HYDROLYSIS OF STARCH

COLOR WITH IODINE

Time(min)	STARCH + SALIVA	STARCH ONLY
2	_____	_____
4	_____	_____
6	_____	_____
8	_____	_____
10	_____	_____
12	_____	_____
14	_____	_____
16	_____	_____
18	_____	_____
20	_____	_____

A-2   RESULTS OF BENEDICT'S TEST:

SAMPLE	COLOR WITH BENEDICT'S REAGENT	IS GLUCOSE PRESENT?
Starch + saliva	_____	_____
Starch only	_____	_____

QUESTIONS:

1. Where does starch digestion begin?

2. What carbohydrate digestion occurs in the small intestine?

3. What are the end products of carbohydrate digestion?

4. Why do we need a digestive process?

EXPERIMENT 43   DIGESTION OF FOODSTUFFS   NAME_____

B. DIGESTION OF TRIACYLGLYCEROLS

B-1 Bile Salts

       MIXTURE   OBSERVATIONS

  Oil and water   _____

                       _____

  Oil, water and bile salts_____

                       _____

B-2 Hydrolysis by Lipase

     TIME(MIN)   pH   VOLUME   0.1 M NaOH(mL)

        0   _____   _____

       20   _____   _____

       40   _____   _____

       60   _____   _____

QUESTIONS:

1. What is the effect of bile salts on an oil and water mixture?

2. What is the function of bile salts in the digestion of fats and
    oils (triacylglycerols)?

3. What products of lipase action would change the pH of a mixture
   containing triacylglycerol?

C. PROTEIN DIGESTION

TEST TUBE

	1 WATER + HCl	2 PEPSIN + HCl	3 PEPSIN ONLY
Egg White Initial Appearance			
Final Appearance			
Has any digestion taken place?			
What caused the digestion process?			

QUESTIONS:

1. Why does a person with a low production of stomach HCl
    have difficulty with protein digestion?

2. What are the products of protein hydrolysis?

EXPERIMENT 44   ENERGY PRODUCTION IN THE LIVING CELL

PURPOSE

1. Observe the reactions of glucose in glycolysis
    and fermentation.

MATERALS

1 package dry yeast
fermentation tubes(5)
(or 5 test tubes and 5 small test tubes that fit inside)
10% glucose solution   Erlenmeyer flasks (125 mL)
10% sucrose solution   thermometer
10% starch solution   methylene blue
ethanol   mineral oil
beakers   temperature bath (37°C)

KEYED OBJECTIVES IN TEXT: 19-1, 19-2, 19-3, 19-4

DISCUSSION OF EXPERIMENT

The amount of food you eat must be sufficient to meet all your
metabolic needs.   One of these needs is the production of energy
required by the cells to do work.   The production of energy is
primarily the function of glucose obtained from the carbohydrates
in your diet.   The extraction of energy from the glucose
molecules is accomplished in a series of enzyme-catalyzed
reactions.

GLYCOLYSIS

In human cells, glucose initially undergoes glycolysis, a series of
reactions in which glucose is converted to pyruvic acid.

$$glucose \longrightarrow 2 \text{ pyruvic acid } + 4 \text{ H}$$

Under aeorbic conditions (oxygen available), a pair of
hydrogen atoms is transported to the electron transport system,
another series of reactions occurs whereby a total of 3 ATP
molecules are produced. These ATP molecules provide energy for the
cell.   The conversion of glucose to pyruvic acid under aeorbic
conditions can provide a total of 6 ATP.

$$glucose \longrightarrow 2 \text{ pyruvic acid } + 6 \text{ ATP}$$

We will observe the conversion of glucose to pyruvic acid
using methylene blue as a hydrogen acceptor as well as an indicator
of the reaction.   The oxidized form of methylene blue is blue,
while its reduced form is colorless.

$$methylene \text{ blue } + 2H \longrightarrow methylene \text{ blue} \cdot 2H$$

blue   colorless

349

Under aerobic conditions, the pyruvic acid would be oxidized further to acetyl CoA and to CO_2 and H_2O via the citric acid cycle. During these additional oxidative steps, pairs of hydrogen atom are removed and transferred to the electron transport chain for additional ATP production. The overall conversion of glucose to CO_2 and H_2O provides a total of 36 ATP for the cell.

$$\text{glucose} + 6\ O_2 \xrightarrow{\text{aerobic}} 6CO_2 + 6H_2O + 36\ \text{ATP}$$

Under anaerobic conditions, cells do not have oxygen available and the electron transport system does not operate. In this case, the pyruvic acid is converted to lactic acid with no further oxidation possible. The anaerobic conversion of glucose to lactic acid provides only a small amount of energy for the cells. (This ATP is produced by direct phosphoryltion only.)

$$\text{glucose} \xrightarrow{\text{anaerobic}} \text{lactic acid} + 2\ \text{ATP}$$

FERMENTATION

In certain cells such as yeast additional enzymes are present to convert pyruvic acid to ethanol and CO_2. In this process called fermentation glucose is converted to ethanol and CO_2. In this experiment, we will observe the fermentation reaction by observing the production of CO_2 bubbles in a mixture of glucose and yeast cells.

$$\text{glucose} \xrightarrow{\text{yeast}} 2\ \text{ethanol} + 2\ CO_2$$

LABORATORY ACTIVITIES

A.   FERMENTATION

Preparation of Yeast Suspension

Prepare a large beaker (400 mL) about half full of water at approximately 37°C for use throughout this experiment. Place a 7-g package of yeast in a 125-mL Erlenmeyer flask. Add 100 mL distilled water and mix. Pour 20 mL of the yeast suspension into a large test tube. Place the flask containing the remaining portion of the yeast suspension in the warm water bath for 10 minutes to bring the temperature of the yeast suspension to 37°C. Do not overheat.

Prepare a boiling water bath with another beaker about 2/3 full of water. Place the test tube containing the 20 mL yeast suspension in the boiling water for 10 min. This will be used as the boiled yeast suspension.

Preparation of Fermentation Tubes

Fill the fermentation tubes with the solutions listed. Place the tubes in a 37°C water bath, tray, water trough or some beakers. Use the warm water (37°C) to cover as much of the fermentation tubes as possible without causing them to tip over. See Figure 44-1.

test tube	solutions
1	10 mL yeast suspension and fill with water
2	10 mL yeast suspension and fill with glucose solution
3	10 mL yeast suspension and fill with starch solution
4	10 mL boiled yeast and fill with glucose solution
5	10 mL yeast and fill with sucrose solution

Figure 44-1   Fermentation tubes.

Using test tubes as fermentation tubes

If fermenation tubes are not available, place the mixtures in 5 test tubes. Place the smaller test tubes upside down in the larges test tubes. Place your hand firmly over the mouth of the test tube and invert. When the small test tube inside has completely filled with the mixture, return the larger test tube to an upright position. See Figure 43-2. Place the tubes in a 37°C water bath and watch for the appearance of bubbles in the small test tubes.

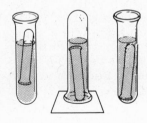

Figure 44-2   Test tubes used as fermentation tubes.

351

Observations of the Fermentation Tubes

Check the fermentation tubes every 20 min during the laboratory period. Add some warm water to the water bath, if necessary, to keep the temperature at about 37°C. Look for the formation of bubbles at the top of the closed tube. If the fermentation tubes have volume markings, record the size of the bubble each time. If not, use a small ruler, to measure the height of the space in the tube occupied by the CO_2 bubble. See Figure 44-3.

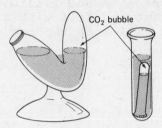

CO₂ bubble

Figure 44-3   Fermentation tubes with CO_2 bubble.

At the end of the testing period, note the odor of each sample. Check the odor of a small sample of ethanol to see if ethanol has been produced.

B.   GLYCOLYSIS(ANAEROBIC CONDITIONS)

Add 2 mL of methylene blue to three test tubes. Prepare the test tubes as follows:

Test Tube	Add
1	5 mL water and 3 mL yeast suspension
2	5 mL glucose and 3 mL yeast suspension
3	5 mL glucose and 3 mL boiled yeast suspension

Add mineral oil dropwise until a 2-3 mm layer forms. This will prevent the oxygen of the air from contact with the reaction mixture. Record the color of the mixtures in the test tubes. Place the test tubes in a beaker of warm water (37°C). Every 20 minutes, observe the color of the mixtures. Record your observations.

EXPERIMENT 44   ENERGY PRODUCTION IN THE LIVING CELL

LABORATORY RECORD   NAME_____
                                     SECTION_____
                                     DATE_____

A. FERMENTATION

OBSERVATIONS
Test tube

Time(min)	1	2	3	4	5
0					
20					
40					
60					
80					
100					
120					

Describe any odor produced by the reaction._____

_____

QUESTIONS:

1. In which test tubes did fermentation occur? Why?

2. In which test tube did fermentation not take place? Why?

3. Write the equation for the fermentation reaction in yeast cells.

EXPERIMENT 44   ENERGY PRODUCTION IN THE LIVING CELL

B. GLYCOLYSIS (ANAEROBIC CONDITIONS)   NAME_____

OBSERVATIONS
Test Tube

TIME(MIN)	1	2	
0			
20			
40			
60			
80			
120			

QUESTIONS:

1. In which test tubes did glycolysis occur? Why?

2. In which test tubes did glycolysis not occur? Why?

3. What are the possible end products of pyruvic acid from glycolysis in aerobic and anaerobic conditions?

EXPERIMENT 45   CHEMICALS IN FOODS AND DRUGS

PURPOSE

1. Describe the chemical components present in food and drugs by reading the labels of containers.
2. Research the function and biological effects of those chemical compounds.

MATERIALS

Processed foods, cosmetics and over-the-counter drugs
Merck Index or the Handbook of Food Additives

DISCUSSION OF EXPERIMENT

By reading the labels on foods and drugs, you will find the chemical names of some of the ingredients used in these products. The properties and functions of some of these compounds can be found in the Merck Index or the Handbook of Food Additives.

LABORATORY ACTIVITIES

1.   Read the labels on six to ten different kinds of products and/or over-the-counter drugs.

2.   Record the ingredients.

3.   Look up at least 10 of the ingredients. Write the formula and structure of each.

4.   Record the function of each and possible dangers.

EXPERIMENT 45   CHEMICALS IN FOODS AND DRUGS

LABORATORY RECORD

Name _____
Date _____
Section _____

Food/Drug	Chemicals Listed	Formula and/or Structure	Function	Dangers

EXPERIMENT 45   CHEMICALS IN FOODS AND DRUGS

LABORATORY RECORD

Name _____
Date _____
Section _____

Food/Drug	Chemicals Listed	Formula and/or Structure	Function	Dangers

EXPERIMENT 46   CHEMISTRY OF URINE

PURPOSE

1. Test urine for pH, specific gravity, and the presence of electrolytes and organic compounds.
2. Test urine for the presence of abnormally occurring compounds of proteins, glucose, and ketone bodies.

MATERIALS

urine samples(normal and pathological)

urinometer	blue cobalt glass
pH paper	6M HNO_3
1% urease	flame-test wire
conc. HCl	1M HAc
litmus paper	0.1M $(NH_4)_2C_2O_4$
test tubes	$(NH_4)_2SO_4(s)$
plastic wrap	5% nitroprusside reagent
beakers	conc. NH_4OH
0.1 M $AgNO_3$	Benedict's reagent
0.1 M $BaCl_2$	6M HCl

ammonium molybdate solution

OPTIONAL: Reagent strips or tablets such as Clinistix, Clinitest, Ketostix, Albustix

DISCUSSION OF EXPERIMENT

Examining a sample of urine can give much information about the processes occurring within the body. The amounts of electrolytes, uric acid, and glucose can all lead to conclusions about the functioning of the kidneys, liver and the general state of health of the individual.

IONS IN URINE

Urine normally consists of about 96% water. The other 4% consists of waste products being eliminated from the cells of the body to maintain proper osmotic pressure, electrolyte levels and pH. Urine normally contains the inorganic ions Cl^-, SO_4^{2-}, PO_4^{3-}, K^+, Na^+, NH_4^+, and Ca^{2+}. Organic components normally found include urea and uric acid. Urea is an end product of protein metabolism and uric acid is an end product of purine metabolism.

$$\overset{\displaystyle O}{\overset{\displaystyle \|}{NH_2-C-NH_2}} \quad \text{urea}$$

The presence of Na^+ and K^+ ion can be determined by flame tests. The present of other electrolytes will also be determined. Uric acid will be detected through the formation of uric acid crystals.

359

EXPERIMENT 46   CHEMISTRY OF URINE

GLUCOSE

Glucose, if present, can be detected by Benedict's test. Glucose
may show up in urine (glucosuria) when high amounts of glucose
accumulate in the blood and the renal threshold is exceeded.
Conditions such as diabetes mellitus and liver damage may be
indicated.

pH AND SPECIFIC GRAVITY OF URINE

Urine usually has a pH around 6.0, although this varies
considerably with diet and activity and can have a range from 4.6
to 8.0 at different times. Urine normally has a light yellow
color derived from pigments formed by the breakdown of bilirubin
obtained from the destruction of red blood cells. The normal
range for specific gravity is 1.005 to 1.030. In this
experiment, you will test the pH of a urine sample, measure its
specific gravity and note its color.

KETONE BODIES

Ketone bodies such as acetone and acetoacetic acid may appear in
the urine when large amounts of fat are metabolized for energy
purposes due to an insufficiency of glucose in the diet or an
inability to utilize glucose as in diabetes mellitus. Ketone
bodies are associated with certain diets (low carbohydrate),
starvation, diabetes mellitus, and liver damage. High levels of
protein(proteinuria) may indicate disease or damage to the
kidneys or urinary tract.

LABORATORY ACTIVITIES

Collection of Urine. Use a clean, dry bottle or a beaker to
collect a sample(50 mL) or urine. A morning sample is
preferable. If you will not be testing the urine sample
immediately, store it in a refrigerator until the laboratory
hour. You will carry out the following urinalysis on your own
urine sample and on a pathological sample prepared by your
instructor.

A.   COLOR, pH AND SPECIFIC GRAVITY

A-1   Obtain 50 mL of your own urine and 50 mL of a pathological
      urine sample. Describe the color of the urine.

A-2   Use pH paper to determine the pH of the urine.

A-3   Determine the specific gravity of the urine sample with a
      urinometer.

B. UREA

Place 5 mL of each urine sample in separate test tubes. Add 2 mL
1% urease solution to each. Let the mixtures stand for 1 hour.
Heat the test tubes <u>gently</u> while you hold a piece of moistened
litmus paper across the top of the test tube. The evolution of
ammonia will turn the paper blue, indicating the presence of urea.
Record results.

$$H_2O \quad + \quad NH_2\overset{\overset{\displaystyle O}{\|}}{C}NH_2 \quad \xrightarrow{\text{urease}} \quad 2NH_3 \quad + \quad CO_2$$
$$\qquad\qquad\qquad\qquad \underset{\text{urea}}{} \qquad\qquad\qquad\qquad \underset{\text{ammonia}}{}$$

C. URIC ACID

Place 25 mL of each urine samples in small, separate beakers. Add
20 drops of conc. HCl to each. **CAUTION: CONC. HCl MUST BE HANDLED
CAREFULLY. USE GOGGLES.** Cover the beaker with plastic wrap. Let
the solutions stand until the next laboratory period. Look for the
appearance of any uric acid crystals. Record results.

D. ELECTROLYTES

D-1 <u>Chloride</u>   Place 30 mL of each urine sample in separate test
     tubes. Add 1 mL 6M HNO_3 and 10 drops 0.1M $AgNO_3$. A white
     precipitate(AgCl) confirms the presence of chloride.

D-2 <u>Sulfate</u>   Place 3 mL of each urine sample in separate test
     tubes.   Add 1 mL 6M HNO_3 and 10 drops of 0.1M $BaCl_2$. A
     white precipitate ($BaSO_4$) confirms the presence of sulfate.

D-3 <u>Phosphate</u>   Place 5 mL of each urine sample in separate
     test tubes. Add 2 mL 6M HNO_3 and 3 mL ammonium molybdate
     solution. Heat gently. A yellow precipitate confirms the
     presence of phosphate.

D-4 <u>Sodium</u> Clean a flame-test wire in 6M HCl and then dip the
     wire in one of the urine specimens. Heat the loop of the
     wire in a flame.   A bright-yellow flame indicates the
     presence of sodium ion.

     <u>Potassium</u> Dip a flame-test wire into one of the specimens
     again. Cover your eye with a blue cobalt glass square, and
     observe the color of the flame through the glass as you heat
     the wire. The appearance of a deep red flame through the
     glass indicates the presence of potassium. Repeat the flame
     tests with each urine sample.

D-5 <u>Calcium</u>   To 3 mL of each urine sample, add 5 drops 1M HAc
     and 2 mL 0.1 M $(NH_4)_2C_2O_4$. A cloudy, white precipitate of
     calcium oxalate confirms the presence of calcium.

361

E. GLUCOSE

Place 8 drops of each urine sample in separate test tubes. Add 5 mL Benedict's reagent to each. Place the test tubes in a boiling water bath for 5 minutes. Record any changes in color. If glucose is present, estimate the amount.

Color with Benedict's	mg%	mg/dL
Blue	0.10	100
Blue-green	0.25	250
Green	0.50	500
Yellow	1.00	1000
Orange	2.00	2000

Commerically availabe test strips such as Clinistix or tablets such as Clinitest may also be used to test for the presence of glucose. Follow directions on the package.

F.   KETONE BODIES

Place 5 mL of each urine sample in separate test tubes. Saturate each with ammonium sulfate, $(NH_4)_2SO_4$. Add 5 drops nitroprusside reagent. **CAUTION: TOXIC.** Tip the test tube and CAREFULLY add 20 drops of concentrated NH_4OH down the side. The presence of a purple ring where the layers meet indicates the presence of ketone bodies.

Commerically available test strips such as Ketostix may also be used to test for the presence of ketone bodies. Follow directions on the package.

G.   PROTEINS

Place 5 mL of each urine sample in separate test tubes. Filter urine first if it is not clear. **CAREFULLY** heat the upper one-half of the liquid to boiling using the Bunsen burner. Add 5 drops 1M HAc. Heat the upper portion for another 1-2 minutes. The formation of a white cloudy precipitate indicates the presence of protein.

Commercially available test strips such as Albustix may also be usd to determine the presence of protein(albumin) in the urine samples.

H.   BAR GRAPH OF ELECTROLYTES

Use graph paper to draw a bar graph of the cations and anions typically found in body fluids. Graph paper can be found at the back of this lab book.

CHEMISTRY OF URINE   NAME_____
                                               SECTION_____
DATE_____

A. COLOR, pH AND SPECIFIC GRAVITY

	Own Urine	Pathological Urine
A-1 Color	_____	_____
A-2 pH	_____	_____
A-3 Specific Gravity	_____	_____

What is your interpretation of the above tests?

B.   UREA

                          Own Urine   Pathological Urine

   Effect on litmus   _____   _____

what is your interpretation of the urea test?

C. URIC ACID

	Own Urine	Pathological Urine
Crystals	_____	_____

What is your interpretation of the uric acid test?

Is uric acid normally found in urine?   Why?

D. ELECTROLYTES

Indicate the presence of electrolytes as follows:
Not present (-)   Present(+)   Strongly Present (++)

	Own Urine	Pathological Urine
D-1 Cl^-	_____	_____
D-2 SO_4^{2-}	_____	_____
D-3 PO_4^{3-}	_____	_____
D-4 Na^+	_____	_____
K^+	_____	_____
D-5 Ca^{2+}	_____	_____

What is your interpretation of the above tests?

EXPERIMENT 46   CHEMISTRY OF URINE   NAME_____

E. GLUCOSE

	Own Urine	Pathological Urine
Benedict's	_____	_____
Estimate of mg%	_____	_____
Estimate of mg/dL	_____	_____
Test paper(_____)	_____	_____
name		

What is your interpretation of results?

When would glucose be found in a urine sample? Why?

F. KETONE BODIES

	Own Urine	Pathological Urine
Color of ring	_____	_____
Test paper(_____)	_____	_____
name		

What is your interpreation of the test for ketone bodies?

When would ketone bodies be found in a urine sample? Why?

G.   PROTEIN

                            Own Urine   Pathological Urine

Appearance   _____   _____

Test paper(_____)   _____   _____
         name

What is your interpretation of the protein test results?

When would protein appear in a urine sample?

G.   BAR GRAPH OF ELECTROLYTES

The following lists the component in a typical urine sample at a pH
of 5.5   Draw a bar graph to represent this data.   Graph paper
may be found at the back of this lab book.

Concentration (mmol/L) of Electrolytes in Urine pH 5.5

Cations		Anions		Nonelectrolytes	
Na^+	170	Cl^-	175	Urea	300
K^+	90	HPO_4^{3-}	45	Creatinine	10
Ca^{2+}	2	SO_4^{2-}	60		
Mg^{2+}	3	Organic acids	45		
NH_4^+	60				

APPENDIX A   OPERATION OF A TRIPLE-BEAM BALANCE

The triple beam balance consists of a weighing pan and a group of
weights each resting on its own beam. You can follow these general
steps for checking the balance and finding the mass of an object.
See Figure A-1.

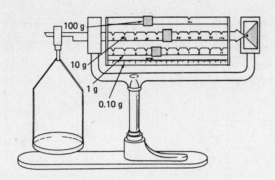

Figure A-1   A Typical Triple Beam Balance.

PROCEDURE FOR FINDING THE MASS OF AN OBJECT

1.   With the weighing pan empty, bring all the weights to the zero
     points on each beam.

2.   Check the swing of the pointer. It should swing the same
     distance above and below the center line. If not, ask your
     instructor to show you how to adjust the balance. When the
     pointer hits the center line with all weights on zer and an
     empty weighing pan, the balance is "zeroed" and ready to use.

3.   Place the object to be weighed on the weighing pan.

4.   Start balancing by moving the largest weight to the right. If
     your balance has notches in the beam be sure the place the
     weight directly in the numbered notch. As you move the
     weight, watch for the pointer to fall <u>below</u> the center
     line(too much weight), then move it back one notch.

5.   Continue to move each of the smaller weights in the same way,
     until the smallest weight(the slide) brings the swing of the
     pointer to the center line. The object and the weights are
     now balanced.

6.   Record the sum of the weights. With most beam balances, you
     will find that the front beam is divided into one-tenth
     gram(0.1 g).   The divisions <u>between</u> the 0.1 g markings are
     read as one-hundredth gram(0.01 g).   Be sure you express mass
     measurement to two figures after the decimal point, even if
     those figures are zero.   (See Figure A-2).

7.   Return all weights to zero.

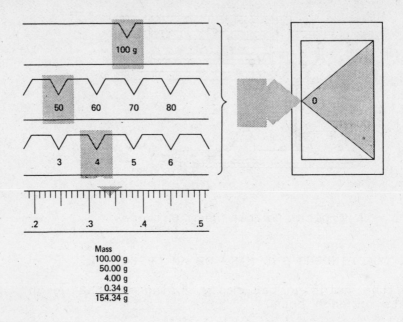

Mass
100.00 g
50.00 g
4.00 g
0.34 g
154.34 g

Figure A-2   Reading the Sum of Weights on a Beam Balance

APPENDIX B   OPERATION OF AN ELECTRIC TOP-LOADING BALANCE

1.   Check that the bubble is centered. See instructor for
     adjustment.

2.   Turn on the balance by the switch in front. See Figure B-1.

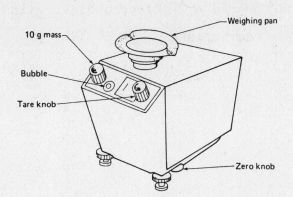

Figure B-1   A Top-Loading Electrical Balance

3.   Check that the **tare** knob is turned as far to the left as it
     will go.

4.   Zero the balance with the **zero knob** on the lower right-hand
     side until the pointer is set at 0.00 g.

5.   Open the cover and place the object on the pan. When weighing
     powders, place them on a preweighed paper; for liquids, use a
     preweighed beaker.

     The **tare** knob can be used when you want to set the mass of a
     container to zero. It is adjusted to 0.00 g with the empty
     container on the pan. Then the mass you obtain is just for
     the substance you place in the container.

6.   Read the tape in the window. If a +++ appears, add 10.00 g by
     turning the **10.00 g mass** knob.

7.   Read the mass of the object to the one-thousandth of a gram
     (0.001 g).

8.   Return the weights (and tare) to zero.

9.   Turn off the machine.

APPENDIX C   CONSTRUCTING A GRAPH

DATA TABLE

Let us suppose that we wanted to measure the distance and time that a bicycle ride traveled. We might take measurements of the distance covered by the rider at certain times. This would give a data table that might look like this.

Distance Covered by a Bicycle Rider with Time

DATA TABLE

Time (hr)	Distance Covered (km)
1	5
3	14
4	20
6	30
7	33
8	40
9	46
10	50

Now we are ready to construct a graph to represent this information.

CONSTRUCTION OF THE GRAPH

1. Draw vertical and horizontal axes on the graph paper. The lines should be set in to leave a margin for numbers and labels. The idea is to fully utilize the graph paper, not just a corner. Place a title at the top of the graph. The title should be derived from the data table. See Figure C-1.

Distance Covered by Bicycle Rider with Time

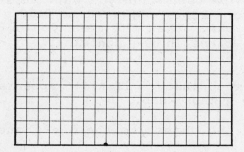

Figure C-1   Axes Drawn On Graph Paper.

2.   Label each axis with the type of measurement and its units. The labels and units are derived from the measurements you took when you prepared the data table. Labels in this graph would be distance (km) on the vertical axis and time(hr) on the horizontal axis. See Figure C-2.

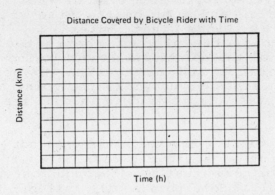

Figure C-2   Labels Placed on Each Axis.

3.   Each axis represents a scale for that measurement. To select a proper scale, observe the lowest and highest numbers in each set of data. The values represented on an axis must be **equally spaced** and fit on the line you have drawn.. The scale cannot exceed the line nor should it cover only a small portion of the line. Utilize as much of the full space as possible.

     For our sample graph, we might set up a scale for distance that starts at 0 km and goes to 60 km. A convenient interval might be 5 or 10 km. Note that you only have to mark a few lines in order to interpret the scale. It gets too crowded with numbers to try and number every line. The time scale uses time intervals of 1 hour to cover the 10 hour time span for the experiment. See Figure C-3.

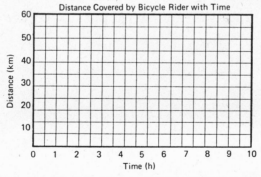

Figure C-3   Marking the Axes with Equal Intervals

PLOTTING THE INFORMATION

4. Now you are ready to plot the data on the graph. Each pair of measurements makes a point on the graph. For example, the rider has covered 20 km at a time of 4 hr. Find 20 km on the distance scale, and 4 hr on the time scale, and imagine perpendicular lines that intersect. That point of intersection is the plot of that set of measurements. Plotting all of the points will begin to show the relationship between distance and time. (You do not need to draw intersecting lines - just find the point of intersection.) Sometimes a small circle is drawn around the point to make it more visible.

5. **Connect** the points. However, this does not mean to jump from point to point in a zig zag fashion. It means drawing a smooth line or curve that goes through most of the points, but perhaps not all. Some points will fall off the line or cruve you draw because there is some error associated with any measurement. Rarely will all points fall on a straight line. See Figure C-4.

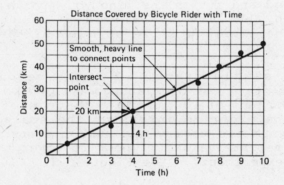

Figure C-4 A Final Graph with Points Connected.

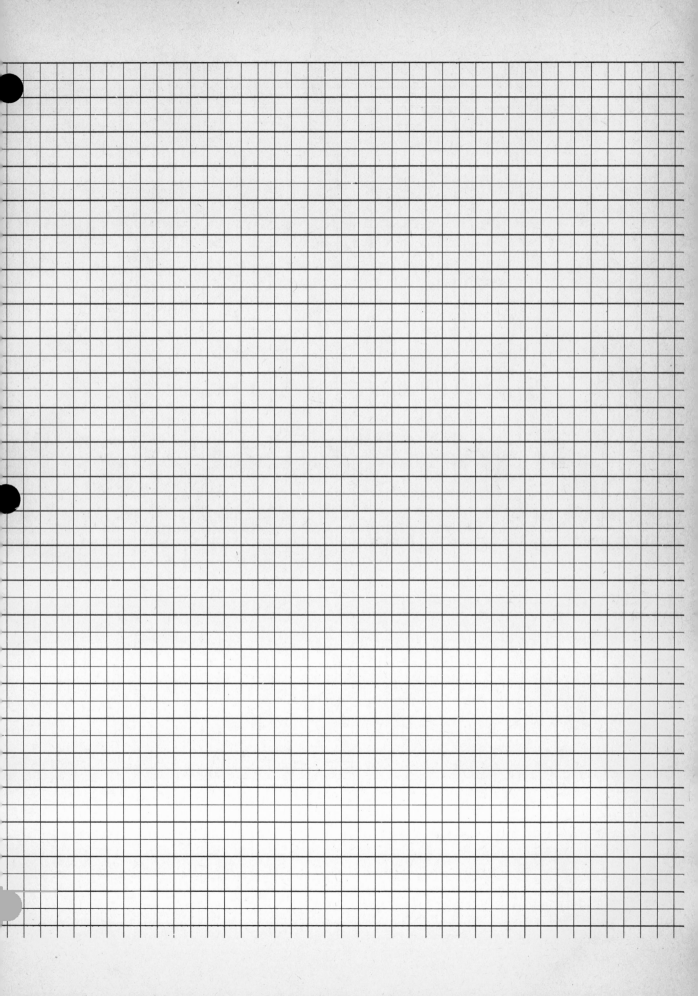

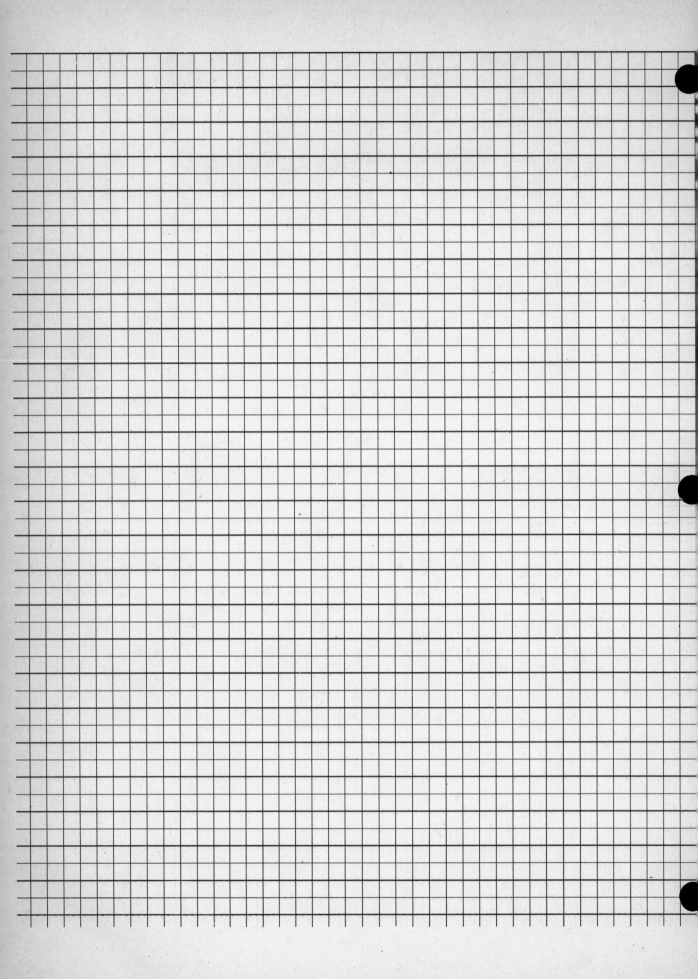

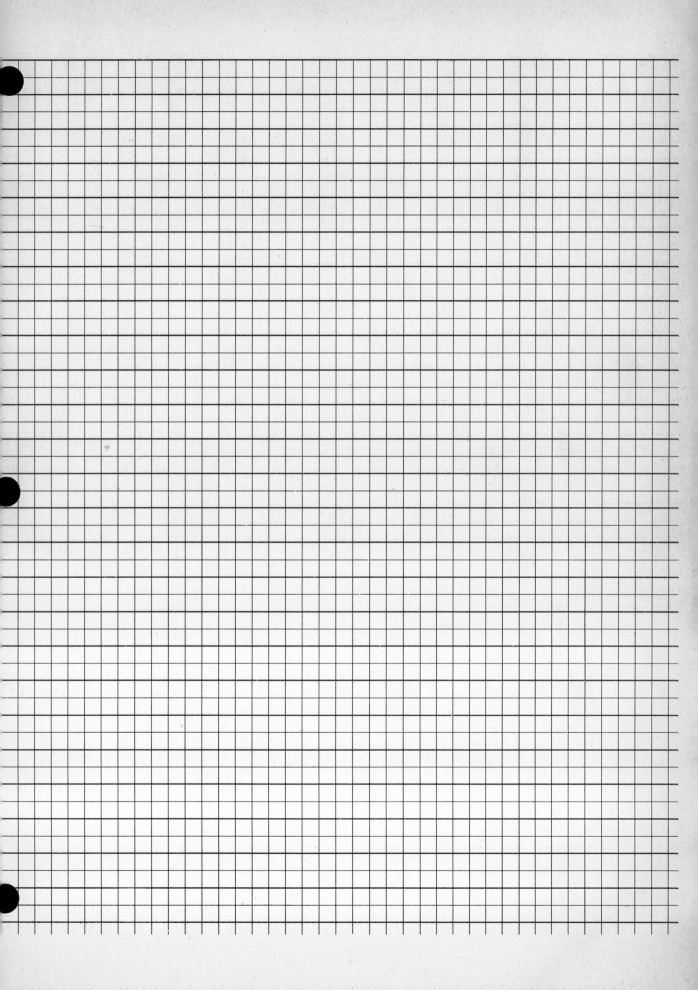

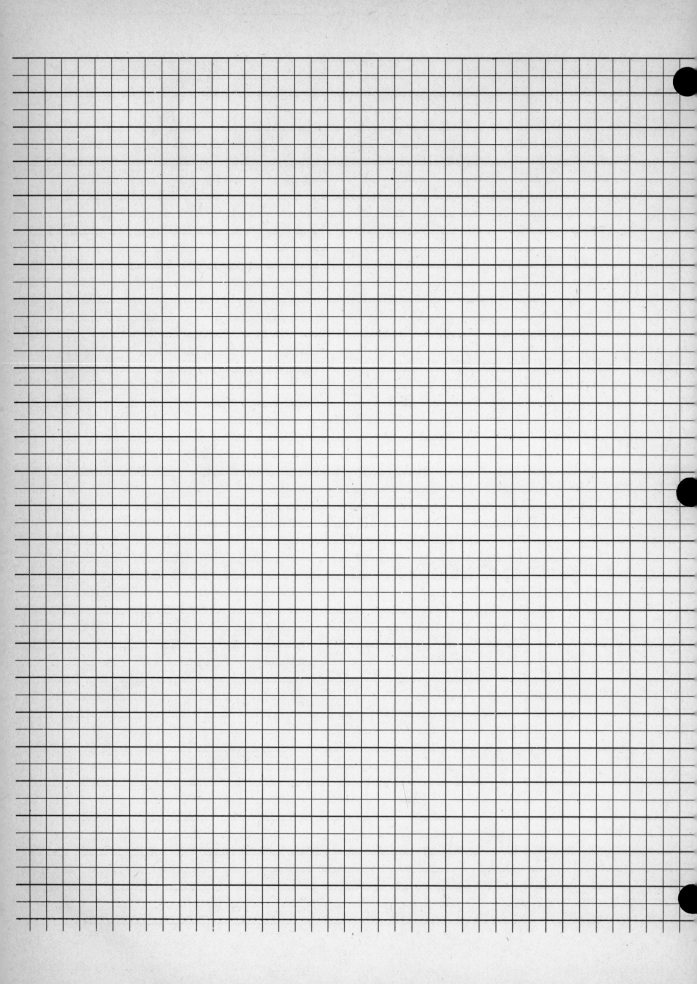

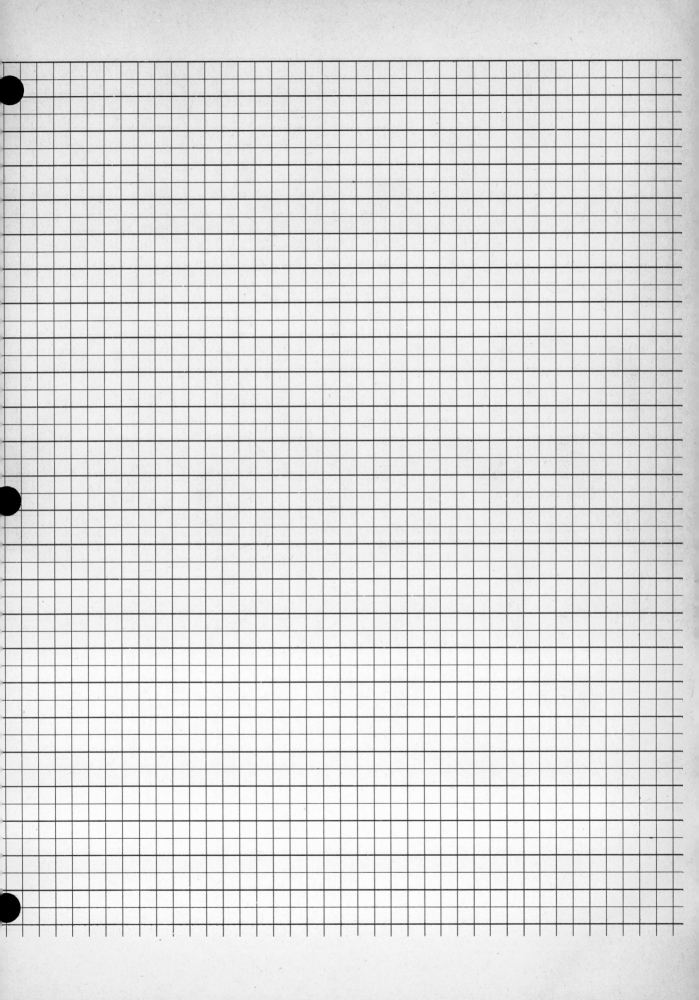

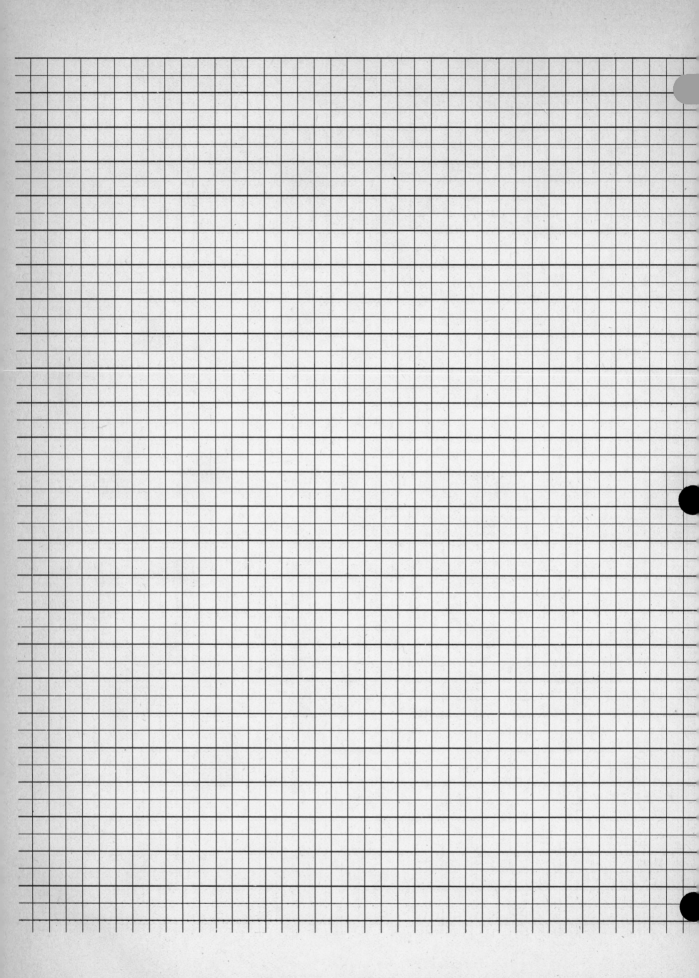

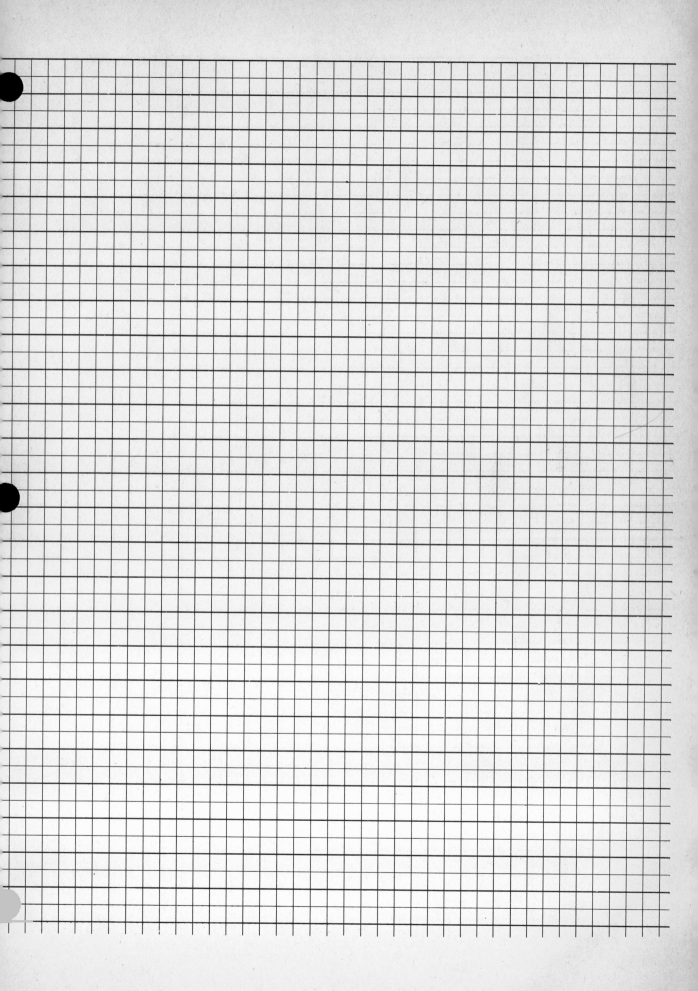

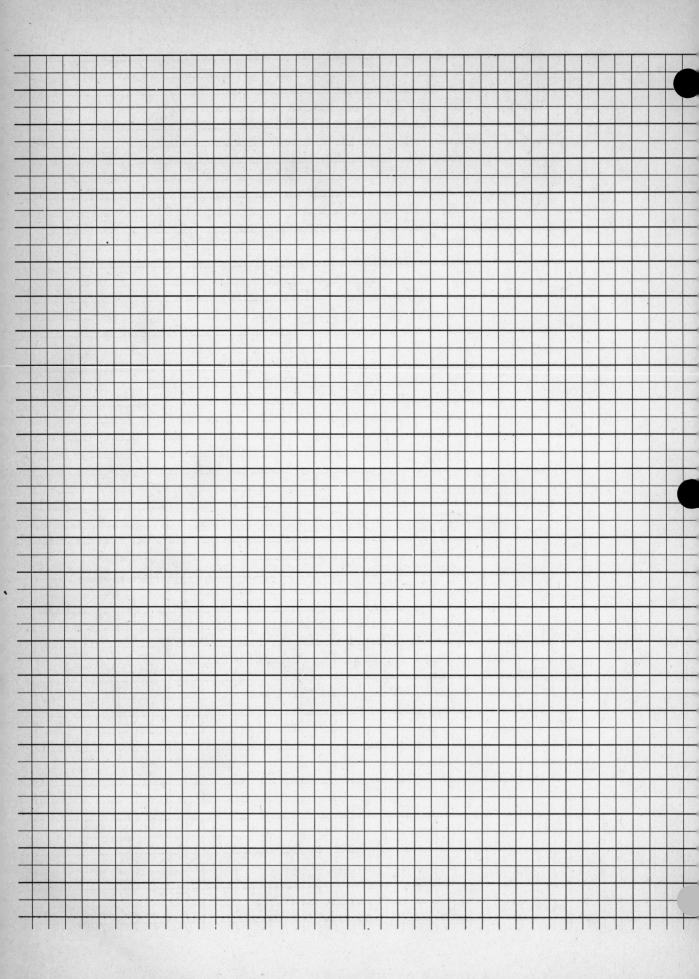